The Science of Answering Questions

A Statistical Approach

First Edition

Jared Studyvin
University of Wyoming

Bassim Hamadeh, CEO and Publisher
Scott De Herrera, Associate Acquisitions Editor
Michelle Piehl, Senior Project Editor
Susana Christie, Senior Developmental Editor
Abbey Hastings, Production Editor
Kim Scott, Freelance Cover Designer
Alexa Lucido, Licensing Manager
Natalie Piccotti, Director of Marketing
Kassie Graves, Senior Vice President, Editorial
Jamie Giganti, Director of Academic Publishing

Printed in the United States of America.

Contents

List of Figures

List of Tables

Chapter 1

Statistics

1.1 Terms

Statistics: The science of making accurate and reliable observations, decisions, and predictions about populations by using information from samples (see Chapter 3 for more information on samples and populations).

Descriptive statistics: Information obtained from a sample to summarize and describe that sample.

Inferential statistics: Information obtained from a sample, taken from a population, to make a prediction about that population.

Question of interest: The question we are trying to answer using descriptive or inferential statistics. Answers to this question are the specific observations, decisions, or predictions about the population. This question is not statistics, but statistics are used to answer this question.

1.2 Finding an Answer

We will use the tools of statistics to answer the question of interest. This is where research begins. Someone has a question they want answered, and the types and complexity of the question can be very broad. A simple example is the question, "Is it going to rain today?" Almost every day we ask ourselves this question. We answer it based on what we wear, even if this all happens subconsciously.

There are several ways to answer a question of interest: guess, use prejudices or opinions, or make an educated guess based on information. For example, the question of interest could be, "Does standardized testing in high school promote better education?" I could answer with, "Yes, it does" or "No, it doesn't." How did I come by those answers? I guessed. I could also answer with, "Yes, it does; having the same standard for all students ensures the quality of the education," or "No, it doesn't; having a single standard doesn't allow for smart students to express their intelligence in their own unique way." How did I come to those answers? Based on my opinions on standardized testing and unique intelligence. The point is there was no information used to come to these answers. If the answer comes from a guess or from an opinion/prejudice, then statistics was not used.

Statistics is the third way of answering the question of interest. Statistics is still a guess, but that guess is based on information. I can collect information from students who went through a system with standardized testing and from students who did not. I can see which students went to college and where. I can measure the college GPA of these students. I can measure their income after 10 years. I can use some information to make an educated guess as to whether standardized testing promotes better education. This is the process of statistics: starting with a question, collecting information, using statistical tools on the information, and coming to an answer based on the results.

1.3 Outline

This text is about the process of reaching an answer through using statistics on collected information. Chapters 2 and 3 are all about the evidence needed to answer the questions of interest. Chapters 5 and 6 are about describing the information collected from the sample. Chapters 7, 8, and 10 have more descriptive statistics and begin the prediction or inferential statistics. Chapters 11 and on are all about either answering different questions of interest or the same question with different types of collected information.

This text is set up to not require the use of software. However, statistical software is very common and is typically used. Chapter 4 is an introduction to a commonly used and freely available choice for statistical software. Several of the chapters have an example section that use software for the analysis.

As alluded to in the terms at the beginning of this chapter and described in more detail in Chapter 3, a sample is not a perfect representation of a population. In other words, predictions from a sample about a population are not perfect. There is uncertainty in this prediction. Statistical probability distributions are used to quantify this uncertainty. Chapter 9 gives an overview of the statistical probability distributions used in this text.

Chapter 2

Variables and Measurement

After a question of interest or research question has been proposed, information needs to be collected to help answer that question. The collected information can be thought of in two parts: what do you want to know and from what group? This chapter is about the what you want to know. Chapter 3 describes the group. What you want to know should match the question of interest. How that information is obtained is the variable and level of measurement.

It is important to select variables that correspond to the question of interest. For example, who is stronger: football players or basketball players? This is a reasonable question of interest. If the variable is measuring the heights of the players, that does not really measure strength. This would be an example of the variable not corresponding to the question of interest.

The issue of matching the variable to the question of interest is typically considered in terms of validity and reliability. Validity assesses the variable for measuring what is desired to be measured. Validity is typically easy to assess for physical measurements (height, width, color, etc.) but is harder to assess for other types of variables such as depression, intelligence, happiness, and so forth. Reliability assesses the variable for being repeatable: Does the variable always measure the same thing every time a measurement is taken? Again, this is easier to assess for physical measurements but harder for other types of variables like depression, intelligence, and happiness.

Assessing the validity and reliability of a variable is beyond the scope of this text. In general, validity and reliability are not really statistical topics. These issues are typically considered within the specific discipline in which the statistical tools are applied. However, statisticians and all scientists/researchers should always be cognizant of validity and reliability when conducting research.

2.1 Terms

Variable: A characteristic of something that can be measured and varies.

Measurement: A means of assigning values (usually numbers) to a variable's possible response.

Outcomes: The possible values that are measured from the variable.

Subject: The person or object that the variable is being measured on. Also called the sample unit.

Varies: Measurements that differ; the outcomes or values are not the same for every subject.

Constant: Measurements that do not vary across subjects; not a variable.

Categorical (nominal): A level of measurement. The outcomes cannot be placed in some logical order; they are just names (nominal) of the possible outcomes.

Ordinal: A level of measurement. The outcomes can be placed in some logical order, but the ordering can only be expressed as more, less, greater, or smaller. The exact difference between the outcomes is not known.

Ratio: A level of measurement. The outcomes are numeric, and the exact difference between outcomes is known. The value of zero is meaningful.

Interval: A level of measurement. The outcomes are numeric and the exact difference between outcomes is known. The value of zero is *not* meaningful.

2.2 What Is a Variable?

A variable is a characteristic of something. Usually these characteristics are of interest. The characteristic must also vary between subjects, which means that different measurement outcomes are possible.

Measurement is the systematic assignment of unique values to a variable. These values can be numerical or nominal in nature. In other words, numbers, names, or a combination of both are assigned to each subject based on the variable.

Varies is a very important part of the word *variable*. If several subjects are measured and the outcome is the same value for all subjects, then this characteristic is called a constant. If the measured outcomes are different (some may be the same, but not all), this measurement varies; this characteristic is a variable.

2.3 Examples

Humans are easy to collect measurements from, since we are all humans. There are lots of great characteristics (variables) to be measured on humans:

1. Height

    (a) Certainly height is a characteristic of humans.

    (b) We can measure a person's height in inches (or feet, meters, etc.).

    (c) Since not everyone is the same height, the measurement values for this variable will certainly vary.

    Conclusion: Height is a variable; possible outcomes are 59", 73", etc.

2. Can you hear?

    (a) Certainly hearing, if a person can hear, is a characteristic of humans.

    (b) We can measure a person's hearing through different tests.

    (c) Since some people can hear and some cannot, the measured outcomes vary.

    Conclusion: Hearing is a variable; possible outcomes are yes or no.

Now let us change the group of people from whom we are interested in obtaining information. Instead of being interested in any human, we are interested only in humans who are completely deaf.

3. Height

    (a) Certainly height is a characteristic of deaf people.

    (b) We can measure a deaf person's height in inches (or feet, meters, etc.).

    (c) Since not every deaf person is the same height, the measurement values for this variable will certainly vary.

    Conclusion: Height is a variable for deaf people.

4. Can you hear?

    (a) Certainly hearing, if a person can hear, is a characteristic of deaf humans.

    (b) We can measure a deaf person's hearing through different tests.

    (c) Every deaf person will have the same outcome from a hearing test.

    Conclusion: Hearing for deaf people is not a variable. It is a constant.

2.4 Levels of Measurement

There are an infinite number of possible measurement systems. All of these can be identified as one of the general types or **levels of measurement**. These levels of measurement are **categorical**, **ordinal**, **ratio**, and **interval**. Knowing the level of measurement will be very useful for the remainder of this book. Important decisions will be made based on the level of measurement of the variable.

2.4.1 Categorical

Categorical variables are also called nominal or qualitative variables. With categorical variables, the outcomes can be differentiated based on categories. Even though we might use numbers to name the categories, the numbers have no quantitative meaning. Examples of categorical variables include the following:

- **Religious affiliation**—Possible outcomes: Christian, Muslim, Jewish, Buddhist, Hindu, and so forth. We cannot order the outcomes. The names are just names of categories.
- **Sex of a subject**—Possible outcomes: female or male. Instead of listing female and male, we could list F and M, a shorthand symbolic notation. Or we could use a numerical shorthand where 1 stands for female and 0 for male. Note, the numbers do not mean anything in a quantitative sense. They only act as names.
- **Eye color**—Possible outcomes: blue, green, brown, and so forth.

2.4.2 Ordinal

With an ordinal variable, the outcomes have a natural ordering. We can tell one outcome is larger (or smaller) than another outcome, but we don't know specifically by how much. Examples of ordinal variables include the following:

- **Places for the best science fair project**—Possible outcomes: first place, second place, and so on.
- **Using the scale provided, respond to the following statement: "The current president of the United States is the best in history."** (This particular type of ordinal variable is called a Likert scale.)

1	2	3	4	5
Strongly Disagree	Disagree	Neutral	Agree	Strongly Agree

Just because we assign a 2 to Disagree and a 1 to Strongly Disagree does *not* mean Disagree is double Strongly Disagree. The outcomes can be placed in order, but we do not know the magnitude difference between the outcomes.

- **What decade you were born in**—Possible outcomes: 60s, 70s, 80s, 90s, and so on.
  A person born in the 80s and a person born in the 90s may not be exactly 10 years apart. If the first person was born in 1989 and the second born in 1993, they are not 10 years apart. Therefore, someone born in the 80s is older than someone born in the 90s, but we do not know exactly how much older.

2.4.3 Ratio

A ratio variable has a natural ordering of the outcomes similar to an ordinal variable. Also we know specifically by how much one outcome is larger (or smaller) than another outcome. The difference in the outcomes is consistent and meaningful. A ratio level of measurement is one that examines variables that are quantitatively related. There are three aspects to a quantitative relationship: (a) **order**, (b) **magnitude**, and (c) a **meaningful zero**. The ordinal level of measurement only addresses order. A ratio level of measurement variable has all three. We can answer questions that require us to know specifically how much larger one number is than another and how much larger a particular number is than zero. The third feature also enables us to compare numbers in a ratio sense. For example, if we have a ratio variable with outcomes 12 and 3, we know from (a) that 12 is larger than 3, from (b) that 12 is nine units large than 3, and from (c) that 12 is four times larger than 3. Examples of ratio variables include the following:

- **Cost of a gallon of gas in dollars and cents**—Possible outcomes: $3.45, $4.75, $8.23.
- **Number of text messages sent**—Possible outcomes: 50, 100, 123.
- **Person's height in inches**—Possible outcomes: 65 in., 70 in., 75 in.

2.4.4 Interval

The interval level of measurement is similar to a ratio level of measurement, except there is no meaningful zero. If we ask an adults how many children they have, they can answer zero, and that means they have no children or that they have an absence of children. This would be a meaningful zero. However, if we measure the outside temperature in degrees Fahrenheit and the outside temperature is zero degrees, this does *not* imply an absence of temperature. This is a variable with a nonmeaningful zero. Temperature still has order and magnitude but no meaningful zero. Examples of interval variables include the following: temperature measured in degrees Fahrenheit or degrees Celsius, GRE score, and SAT score.

2.5 Using Variables in Statistical Calculations

Knowing and understanding the different levels of measurement is very important for interpretations and drawing conclusions from statistical analyses. The mathematics of statistical calculations can, however, generally handle only two of the levels of measurement: categorical and ratio. There are some statistical analyses suited for an ordinal variable, but these are less common. For this reason, in most statistical analyses (calculations) the variable is either treated as categorical or ratio. If the variable is ratio or categorical the choice is easy, but what should be done for ordinal and interval variables? For the remainder of this text some rules will be developed. These rules will be thought of as absolutes in this text, but they are not absolutes in the real world doing real analyses. The purpose of these rules for the level of measurement is to make the material simpler and consistent in decision-making. These rules are really a convention for this text, and other sources use different conventions. Here are the rules:

Ratio variable: Treat it as a ratio variable.

Categorical variable: Treat it as a categorical variable.

Interval variable: Treat it as a ratio variable. In fact, for the remainder of the text, interval variables will not be distinguished. They will be labeled as ratio variables.

Ordinal variable with five or more outcomes: Treat it as a ratio variable.

Ordinal variable with four or fewer outcomes: Treat it as a categorical variable.

These choices are not arbitrary. Experience in conducting statistical analyses has shown that treating an ordinal variable with five or more outcomes as ratio works well. Outside this text, a researcher should not blindly follow these rules but decide what is best in each specific situation.

Chapter 3

Populations, Samples, and Sampling

Sampling is a small piece of the discipline of statistics, but it does represent a rather critical piece. The act of collecting the sample is the small piece, and using the information from the sample to answer the questions of interest is the vast majority of the discipline. There are two essential elements to generating good information. The first is an issue with measurement, having the appropriate level of measurement (see Chapter 2). The other issue is sampling, the topic of this chapter. For observations, decisions, and predictions to be any good, they must be based on good information. Basically, sampling is the science of selecting subjects as a surrogate for the population.

3.1 Terms

Population: The complete collection of subjects that we wish to study. (We want the answer to our question of interest to apply to this whole collection of subjects.)

Sample frame: The complete list of the entire population.

Sample unit: Any single subject from the population.

Sample: The collection of the sample units (subjects) that were selected.

Sample size (n): The number of subjects from the population who are in the sample.

Census: A sample containing all the sample units from the population.

Scores: Observed outcomes of the variable from the sample units in the sample.

Enumeration: The complete listing of the scores for the entire sample.

Statistic: Any information obtained from the sample.

Parameter: Any information obtained from the population.

Sampling: The science of selecting a subset of sample units from the population.

Estimate: The use of a statistic to make an educated guess about a parameter (we use statistics to make an estimate of the population parameter).

Random: Every subject of the population has an equal chance of being selected in the sample and without bias.

Bias: Systematic error or systematic mistakes.

Simple random sampling: A method of sampling where each subject of the population has an equal chance of being selected and the subjects selected will give an unbiased reflection of the whole population (often denoted as SRS).

Convenience sampling: Simply not a simple random sample; the sample has a biased reflection of the population or not every subject has an equal chance of being selected.

3.2 Examples

1. What is the average income of governors in the United States?

   Population: All the governors in the United States.

   Sample frame: A complete list of all 50 governors.

   > **Note**: For the population to be the 50 governors, we do not have to know them to identify them as a population. However, we need to know each of them by some identification to create a sample frame.

   Sample unit: Any one of the 50 governors.

   > **Note**: Even though the word *sample* is used in "sample frame" and "sample unit," these terms apply to the population.

   Sample: Any subset of the 50 governors, for example, any 15 of the 50 governors. The number 15 is the **sample size**, which is decided by the researcher conducting the study.

Census: All 50 governors are in the sample. In practice this is almost never done.

Variable: Yearly income.

Outcomes: Dollar amounts.

Level of Measurement: Ratio.

Parameter: If we collect the income of all 50 governors, the average of those 50 income values is the parameter.

Statistic: If we collect 15 governors' income, the average of those 15 incomes is the statistic.

Estimate: When we use the average from the 15 governors' income to predict (estimate) the average income for all 50 governors.

Sampling: The science of how to select the governors who will be used to collect income values. There are two sampling methods considered in this text: random and convenient.

Simple random sampling: After a sample frame is created of the 50 governors, randomly select 15 governors off the list to be in the sample. This means that each governor has an equal chance of being selected into the sample. Because of the random nature of the sample, the sample is considered representative of the population. This makes the statistic (average income of the 15 governors) a better estimate of the parameter (average income of all 50 governors).

Convenient sampling: Selecting the 15 nearest states to me as the governors to be in the sample. In this case not every governor has an equal chance of being in the sample. Another example is picking the first 15 governors on the list to be in the sample. Convenient sampling is much easier than simple random sampling, which is why it is done a lot. Because of the nonrandom nature of convenient sampling, we are not guaranteed the sample will represent the population. This makes the statistic a poor estimate of the parameter.

2. What is the average GPA of undergraduate students at the local university?

Population: The collection of all undergraduate students at the local university.

Sample frame: A complete listing of all undergraduate students at the local university.

Sample unit: A single undergraduate student.

Sample: Any 40 undergraduate students or any 123 undergraduate students. Any subset of the undergraduate students. If we selected all the undergraduate students(i.e. select the whole population) that would be a **census**.

Variable: Grade point average.

Outcomes: The scale of 0.0 to 4.0, values like 3.7, 2.4, and so forth.

Level of measurement: Ratio.

Parameter: The average GPA of all undergraduate students at the local university.

Statistic: The average GPA of the 10 sampled undergraduate students.

Estimate: When we use the average GPA of the 10 undergraduate students to estimate the average GPA of all undergraduate students at the local university.

Simple random sampling: We list all of the undergraduates' names and randomly select 40 of them.

Convenient sampling: Any way of selecting 40 students so that all the students do not have an equal chance of being selected.

3.3 Sample Size

We know that random is better than nonrandom samples. The next question always is, what sample size should we use? The best answer is, it depends on several different things; a class on sampling will address these issues in greater detail. For the purposes in this book, there are three principles of sampling:

- Random is better than nonrandom.
- Larger sample size is better.
- More samples are better.

Let's use an example to illustrate these principles.

Question: How many students are typically in a fourth-grade class in Wyoming?
How would you answer this question? We need to take a sample. There are 53 school districts in Wyoming.

Possible solution: Randomly select seven school districts. Next, randomly select three fourth-grade classes out of each district and count the number of students. We have a random sample, but how would more samples be taken to get a larger sample size?

Increase samples: Randomly select 25 school districts. This would increase the number of samples.

Increase sample size: Randomly select six fourth-grade classes in every district chosen. This would increase the sample size.

The number of fourth-grade classes is the sample size. The number of districts is the number of samples. It is true that increasing the number of samples (number of school districts) does increase the sample size, but indirectly. To directly increase the sample size, more classes need to be selected.

3.4 Independent Samples

Later, some hypothesis testing situations will be discussed that require two samples (Section 11.6). These test situations require that the two samples be independent of each other. This means that neither samples have the same subjects and are unrelated. What does "unrelated" mean? Basically, if there is a way to determine the samples are related, then they are not unrelated. If nothing can be determined that demonstrates the samples are related, the samples are deemed unrelated. Here are some examples:

1. If sample 1 is mothers and sample 2 is their children, then the samples are related, so the samples are not independent.
2. Sample 1 is women and sample 2 is their brothers. Since they are siblings, the samples are related.
3. Sample 1 is male students and sample 2 is female students. There is nothing in how the samples were selected to determine a relationship between the samples. It is possible there are siblings in the samples, or some other relationship, but the samples were selected in a random fashion and separate from each other. Thus, these two samples would be considered independent.

Chapter 4

Statistical Software

There are many different statistical software packages available such as `R`, Julia, SAS, SPSS, Minitab, Python, and even Microsoft Excel, which can do a few basic statistical calculations. There are many things to consider when selecting a statistical software package. The two main considerations are the cost and ease of use. Programs like SPSS and Minitab are "point-and-click" programs, making them easy to use, but the cost of these programs is much higher. The program `R`, on the other hand, is open source (free) but requires "coding," so it is a little more difficult to use. We will use the `R` statistical programing language.

This text was written so that the reader is not required to learn or use `R`. If you do not wish to learn `R`, skip this chapter and all of the sections titled Examples With `R` in subsequent chapters.

4.1 Terms

Variable: A named placedholder for a value that can change depending on the information passed to the computer program. This is different from the variable term in Chapter 2. This is the computer science definition of a variable.

Logical operators: A comparison between two things, the result of which is either true or false.

Data type: The type or class of the variable. A number and a letter are different types or classes of information that can be stored in a variable.

Data structure: A means of organizing multiple variable values into a meaningful pattern.

Function: Previously written code for a specific task that has been assigned a name. The named function can be used to accomplish the specific task.

4.2 Installing R

The website to download R changes, and the best way to find it is to do an internet search of "download R for Windows," "download R for macOS," or "download R for Linux," depending on your computer's operating system (OS). The R software comes with a graphical user interface (GUI). Figure 4.1 shows the default R GUI. Some people prefer to use a different GUI to edit and run R commands. This chapter will not talk about other possible GUIs to use with R.

4.3 Getting Started With R

The R statistical software is what is called a scripting language. Figure 4.1 shows a screenshot of the R GUI. On the left side is the R console. The greater-than (>) character is the command prompt, and commands (code) can be typed directly into the console. On the right side is a blank R script or .R file. These text files are designed to be a place to write, edit, and store R code. It is recommended to write and store your R code in an R script so that you do not need to retype all the commands each time. How to submit commands from the R script to the console depends on which GUI you are using. Typically, pressing the Control key and the R key or the Control key and the Enter key will submit a command from the script to the console.

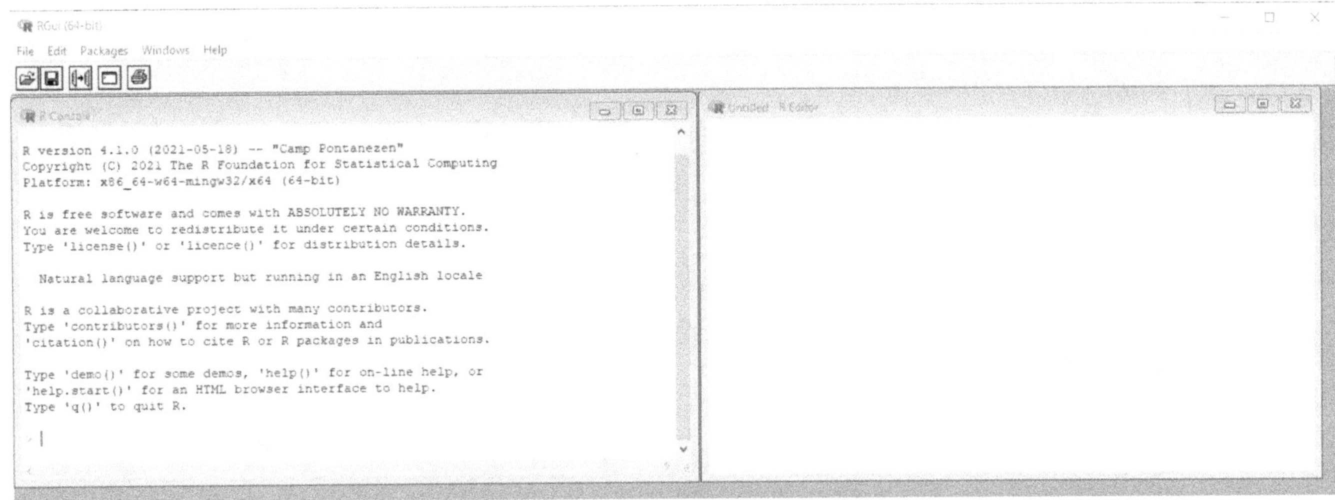

Figure 4.1: Screenshot of graphical user inference when R is installed.

4.4 Arithmetic and Logical Operators

The basic arithmetic operations in R code are shown. The symbol # is used to make comments in R code. Anything after the # symbol is ignored. Remember that powers (^) have precedent over multiplication and division, and these have precedent over addition and subtraction.

```
> ## This is how to make a comment within the codes
> 3+4 # addition
[1] 7
> 5-8 # subtraction
[1] -3
> 16/5 # division
[1] 3.2
> 43*74 # multiplication
[1] 3182
> 4^3 ## powers, 4 to the power 3
[1] 64
> 4 -5^2 ## remember the order of operations
[1] -21
> (4-5)^2 ## using parentheses to control the order of operations
[1] 1
```

Logical operators are a comparison between two things, and the result is either true or false. Here are some examples of logical operators:

```
> ## greater than is the same as greater than or equal to
> 4 > 5 # greater than
[1] FALSE
> 5 > 5 # 5 is not greater than 5
[1] FALSE
> 5 >= 5 # greater than or equal to
[1] TRUE
```

```
> 4 < 5 # less than
[1] TRUE
> 4 <= 5 # less than or equal to
[1] TRUE
> 4 < 4 # FALSE
[1] FALSE
> 4 <= 4 # TRUE
[1] TRUE
> 4 == 5 # test if equal
[1] FALSE
> 4 == 4 # TRUE
[1] TRUE
> 4 != 5 # test for not equal
[1] TRUE
```

It is possible to assess multiple logical statements with "or" and "and" logical operators. Here are some examples:

```
> ## & is the "and" operator
> 4 > 5 & 6 < 7 # both have to be true for the whole to be true
[1] FALSE
> 4 < 5 & 6 < 7 # TRUE, both are true
[1] TRUE
> ## | (called pipe) is the "or" operator
> 4 > 5 | 6 < 7 # only one has to be true for the whole to be true
[1] TRUE
> 4 < 5 | 6 < 7 # both are true so the whole is true
[1] TRUE
> ## evaluation is done left to right
> 5 < 2 | 4 > 1 & 5 > 10
[1] FALSE
```

4.5 Data Types and Variables

One of the powerful things about using a computer (and thus coding) is the ability to assign and use variables. Unfortunately, the definition of the word "variable" here is different from the the definition used in Chapter 2. In software a variable is more like a variable in mathematics: a named placeholder for a particular value. In `R` the variable name can be anything as long as it starts with a letter. Numbers and some special characters can be used in a variable's name. However, it is recommended to use mostly letters and occasionally numbers. It should be noted that `R` is case sensitive; the variable `varOne` is not the same as the variable `VARONE`. Some of the basic types of data a variable in `R` can be are integers, numeric (decimal) values, strings (or characters), and logicals. These basic data types can be combined into more complex data structures.

4.5.1 Numeric and Integer Variables

Here are some examples of the numeric and integers:

```
> varOne <- 4 ## assign the variable varOne the value of 4
> varOne
[1] 4
> varTwo <- 3.4
> ## all of the previous arithmetic and logical operations can be done on the variables
> varOne + varTwo
[1] 7.4
> varOne > 8
[1] FALSE
> varOne != varTwo
[1] TRUE
> varOne^varTwo
[1] 111.4305
```

4.5.2 String Variables

Here are some examples of strings. The use of a double quote or single quote works for assigning string variables. Arithmetic does not work on strings, but some of the logical operators do work.

```
> stringOne <- 'hello'
> stringOne
[1] "hello"
> stringTwo <- 'hello'
> stringThree <- 'good bye'
> ## logical operators
> stringOne > stringTwo
[1] FALSE
> stringOne >= stringTwo
[1] TRUE
> stringOne < stringThree
[1] FALSE
> stringOne > stringThree ## using alphabetical assessment
[1] TRUE
> stringOne == stringTwo
[1] TRUE
> stringOne != stringTwo
[1] FALSE
```

4.5.3 Logical Variables

Logical variables take on the value of true or false. Arithmetic operations work on logical variables because true is treated like a 1 and false is treated like a 0:

```
> logicOne <- TRUE # must be capitalized
> logicOne
[1] TRUE
> logicTwo <- FALSE # must be capitalized
> logicOne * 3
[1] 3
> logicTwo * 4
```

```
[1] 0
> logicTwo & logicOne
[1] FALSE
> logicOne | logicTwo
[1] TRUE
> logicOne + logicTwo
[1] 1
```

4.5.4 Scientific Notation

Really big or small numeric values in R are presented in scientific notation:

```
> 0.000000000000001234
[1] 1.234e-15
```

The e stands for 10 raised to some power. What is provided in R is the same as 1.234×10^{-15}.

```
> 5678000000000000000000000000000000000
[1] 5.678e+35
```

This is the same as 5.678×10^{35}.

4.6 Data Structures and Functions

If a variable has more than one value, it is a data structure. The most common data structures in R are vectors, matrices, arrays, lists, and data frames. The only way to create a data structure in R is to use a function. Therefore, functions must be described first. A function is like a cooking recipes: The function's internal coding is the set of instructions or the recipes to follow. The function can only perform the "recipe" that has been coded for that function. To use a function, type the function's name, then an open parenthesis followed by the argument values and then a close parenthesis. The arguments are the ingredients to be used in the recipe. Most functions have a return statement: the value or data structure to return. These return values can be assigned to new variables. When R is first installed and opened, there are several functions already loaded and available to the user. Here some examples of functions and their use:

```
> varOne <- 4.5
> varTwo <- TRUE
> ## call the function is.logical give the argument named x the value of the variable varOne
> is.logical(x=varOne) ## is varOne a logical variable? FALSE
[1] FALSE
> is.logical(x=varTwo)
[1] TRUE
> log(x=varOne) ## natural log of varOne
[1] 1.504077
> log(x=varOne,base=2) # log base 2 of varOne
[1] 2.169925
```

A function argument can have a default value. `log(x=varOne)` did not specify the `base` argument because the default value for the `base` argument is to use Euler's number as the base, which produces the natural log. There are lots of mathematical functions already built into `R`. A quick internet search should find the exact name of the function that is needed.

There are some variables already assigned in `R` whenever `R` is started. A nice one to remember is π, from the equation for the area of a circle:

```
> pi
[1] 3.141593
```

4.6.1 Vectors

A vector in `R` is several individual values concatenated. All of the values must be of the same data type. Strings and numbers cannot be concatenated:

```
> ## c is function to concatenate several values together, the values are separated by a comma
> vecOne <- c(3,4,58,12)
> vecOne
[1]  3  4 58 12
```

Each value within a data structure is called an element. To access an element or a subset of elements, use the square brackets with the variable name:

```
> vecOne[4] # access the fourth element
[1] 12
> vecOne[c(1,3)] # access the first and third elements
[1]  3 58
```

There are also some nice functions to create a vector with a particular format:

```
> 1:4 # integers from:to
[1] 1 2 3 4
> 5:-3 # works in both directions
[1]  5  4  3  2  1  0 -1 -2 -3
> seq(from=2,to=3,by=.1)
 [1] 2.0 2.1 2.2 2.3 2.4 2.5 2.6 2.7 2.8 2.9 3.0
```

Vectors can also be made up of strings or logical values:

```
> vecLogical <- c(TRUE, FALSE, TRUE)
> vecLogical
[1]  TRUE FALSE  TRUE
> vecString <- c('hello,','how', 'are', 'you?')
> vecString
[1] "hello," "how"    "are"    "you?"
> ## vectors already in R
> letters
 [1] "a" "b" "c" "d" "e" "f" "g" "h" "i" "j" "k" "l" "m" "n" "o" "p" "q" "r" "s"
[20] "t" "u" "v" "w" "x" "y" "z"
> LETTERS
 [1] "A" "B" "C" "D" "E" "F" "G" "H" "I" "J" "K" "L" "M" "N" "O" "P" "Q" "R" "S"
[20] "T" "U" "V" "W" "X" "Y" "Z"
> month.name
 [1] "January"   "February"  "March"     "April"     "May"       "June"
 [7] "July"      "August"    "September" "October"   "November"  "December"
```

```
> month.abb
 [1] "Jan" "Feb" "Mar" "Apr" "May" "Jun" "Jul" "Aug" "Sep" "Oct" "Nov" "Dec"
```

If vector has multiple data types, it will be converted to a single data type:

```
> c(2,29,'hello') # make everything a string
[1] "2"     "29"    "hello"
> c(2, TRUE, FALSE) # make everything numeric
[1] 2 1 0
> c(2, 'hello', TRUE, FALSE) # make everything a string
[1] "2"     "hello" "TRUE"  "FALSE"
```

4.6.2 Factor

A vector in R could be all of the outcomes from a statistical variable. If 10 people were asked their height (in inches), those 10 values could be stored in a vector. The same thing is true for a categorical variable. Those same 10 people could be asked, What is your favorite color: red, green, or blue?" The vectors could be like this:

```
> height <- c(70,71,69,65,73,67,64,70,68,66)
> favoriteColor <- c('blue','green','green','blue','blue','green','blue','green','green','blue')
> height[4] # fourth person's height
[1] 65
> favoriteColor[4] # fourth person's favorite color
[1] "blue"
```

What if these same 10 people were asked, "What is your blood pressure: high, medium, or low?" This is an ordinal variable. In R an ordinal variable is called a factor:

```
> bloodPressure <- factor(x=c('low','medium','medium','low','low','high','low','medium',
                              'low','high'),levels=c('low','medium','high'),ordered=TRUE)
> bloodPressure
 [1] low    medium medium low    low    high   low    medium low    high
Levels: low < medium < high
```

Now the vector bloodPressure is an ordinal variable.

4.6.3 Matrix and Array

A matrix in R is several vectors combined. This can be done in several different ways:

```
> matrixOne <- matrix(data=1:15,nrow=3,ncol=5,byrow=FALSE)
> matrixOne
     [,1] [,2] [,3] [,4] [,5]
[1,]    1    4    7   10   13
[2,]    2    5    8   11   14
[3,]    3    6    9   12   15
> matrix(data=1:15,nrow=3,ncol=5,byrow=TRUE)
     [,1] [,2] [,3] [,4] [,5]
[1,]    1    2    3    4    5
[2,]    6    7    8    9   10
[3,]   11   12   13   14   15
> matrix(data=1:15,nrow=5,ncol=3,byrow=FALSE)
     [,1] [,2] [,3]
[1,]    1    6   11
[2,]    2    7   12
[3,]    3    8   13
[4,]    4    9   14
[5,]    5   10   15
> matrix(data=1:15,nrow=5,ncol=3,byrow=TRUE)
     [,1] [,2] [,3]
[1,]    1    2    3
[2,]    4    5    6
[3,]    7    8    9
[4,]   10   11   12
[5,]   13   14   15
> cbind(1:5,6:10,11:15)
     [,1] [,2] [,3]
[1,]    1    6   11
```

```
[2,]    2    7   12
[3,]    3    8   13
[4,]    4    9   14
[5,]    5   10   15
> rbind(1:5,6:10,11:15)
     [,1] [,2] [,3] [,4] [,5]
[1,]    1    2    3    4    5
[2,]    6    7    8    9   10
[3,]   11   12   13   14   15
```

A matrix can also have strings or logical values:

```
> matrix(data=letters[1:9],ncol=3,nrow=3)
     [,1] [,2] [,3]
[1,] "a"  "d"  "g"
[2,] "b"  "e"  "h"
[3,] "c"  "f"  "i"
> matrix(data=c(TRUE,TRUE,FALSE,FALSE),ncol=2,nrow=2)
     [,1]  [,2]
[1,] TRUE FALSE
[2,] TRUE FALSE
```

A matrix is two dimensional; it has rows and columns. The square brackets are used to access elements from the matrix, but the row and the column have to be specified:

```
> matrixOne[1,3] # first row, third column
[1] 7
> matrixOne[2,] # second row
[1]  2  5  8 11 14
> matrixOne[,4] # fourth column
[1] 10 11 12
> matrixOne[1:2,3:4] # first and second rows and third and fourth columns
```

```
     [,1] [,2]
[1,]    7   10
[2,]    8   11
```

Notice that, depending on what subset of elements is requested, a vector or matrix is returned.

Matrices and vectors are special cases of an array. A matrix has two dimensions, and a vector has one dimension. An array can have as many dimensions as the user wants. However, it can only be printed (or displayed) in the console in two dimensions:

```
> arrayOne <- array(data=1:12,dim=c(2,2,3))
> arrayOne
, , 1

     [,1] [,2]
[1,]    1    3
[2,]    2    4

, , 2

     [,1] [,2]
[1,]    5    7
[2,]    6    8

, , 3

     [,1] [,2]
[1,]    9   11
[2,]   10   12
> arrayOne[1,2,3] # access elements in the same way
[1] 11
```

4.6.4 Data Frame

The data frame data structure is most often used for statistical analysis. In Section 4.6.2, there was an example of 10 people and three variables. The data frame allows the variable outcomes to be connected by each person:

```
> dataFrameOne <- data.frame(heightInches=height,color=favoriteColor,bp=bloodPressure)
> dataFrameOne
   heightInches color     bp
1            70  blue    low
2            71 green medium
3            69 green medium
4            65  blue    low
5            73  blue    low
6            67 green   high
7            64  blue    low
8            70 green medium
9            68 green    low
10           66  blue   high
```

This code creates a data frame with a column named `heightInches` from the vector `height`, a column named `color` from the vector `favoriteColor`, and a column named `bp` from the vector `bloodPressure`. A column of a data frame has to be the same data type, but columns do not need to be the same data type. A data frame is two-dimensional, like a matrix, but now the dimensions have meaning. The rows represent each sample unit, and the columns represent each variable measured on the sample units:

```
> dataFrameOne[3,] # outcomes of the three variable for the third person in the sample
  heightInches color     bp
3           69 green medium
> dataFrameOne[,1] # first column or heightInches variable
 [1] 70 71 69 65 73 67 64 70 68 66
> dataFrameOne[,'heightInches'] # access with the column name
 [1] 70 71 69 65 73 67 64 70 68 66
```

```
> dataFrameOne$heightInches # another way to access with the column name
 [1] 70 71 69 65 73 67 64 70 68 66
> ## measured a new variable and add it to the data frame
> dataFrameOne$favoritePet <- c('cat','dog','dog','dog','cat','dog','cat','dog','cat','cat')
```

There are several nice functions for working with a data frame. Here are a few of them:

```
> nrow(x=dataFrameOne) # number of rows or sample size
[1] 10
> head(x=dataFrameOne) # view first 6 rows
  heightInches color     bp favoritePet
1           70  blue    low         cat
2           71 green medium         dog
3           69 green medium         dog
4           65  blue    low         dog
5           73  blue    low         cat
6           67 green   high         dog
> tail(x=dataFrameOne) # view bottom 6 rows
   heightInches color     bp favoritePet
5            73  blue    low         cat
6            67 green   high         dog
7            64  blue    low         cat
8            70 green medium         dog
9            68 green    low         cat
10           66  blue   high         cat
> names(x=dataFrameOne) # column names
[1] "heightInches" "color"        "bp"           "favoritePet"
> dim(x=dataFrameOne) # number rows and columns, this works for matrix and array
[1] 10  4
> str(object=dataFrameOne) # data type of each column
'data.frame':        10 obs. of  4 variables:
 $ heightInches: num  70 71 69 65 73 67 64 70 68 66
```

```
$ color       : chr  "blue" "green" "green" "blue" ...
$ bp          : Ord.factor w/ 3 levels "low"<"medium"<..: 1 2 2 1 1 3 1 2 1 3
$ favoritePet : chr  "cat" "dog" "dog" "dog" ...
```

4.6.5 List

A list in R is like a vector, but each element can be a different data structure. The list elements can have names or not. Here is an example of a named list:

```
> listOne <- list(string='hello',number=4.5,dataFrame=dataFrameOne,matrix=matrixOne)
> listOne[1:2] # first two elements but the result is still a list
$string
[1] "hello"


$number
[1] 4.5
> listOne['matrix'] # result is still a list
$matrix
     [,1] [,2] [,3] [,4] [,5]
[1,]    1    4    7   10   13
[2,]    2    5    8   11   14
[3,]    3    6    9   12   15
> listOne[['matrix']] # result is a matrix
     [,1] [,2] [,3] [,4] [,5]
[1,]    1    4    7   10   13
[2,]    2    5    8   11   14
[3,]    3    6    9   12   15
```

Here is a helpful function to determine the data structure or type of a variable:

```
> class(stringOne)
[1] "character"
> class(listOne)
```

```
[1] "list"
> class(listOne['matrix'])
[1] "list"
> class(listOne[['matrix']])
[1] "matrix" "array"
> class(dataFrameOne)
[1] "data.frame"
```

4.7 Writing Functions

There are times when writing a function in R is required to complete an analysis. The topic and best practices for function writing are extensive. The purpose here is to give a very brief introduction to using functions.

A function is made up of three basic parts: name, arguments, and body. A cooking analogy works well here. The function name can be likened to the name of the food dish that is desired. The function arguments are the ingredients for the dish. The function body is the recipe to follow to get the desired dish.

The following function example code creates a new function called `mySum`. The desired result is a sum based on the function's name. The function takes two arguments, `x` and `y`, which are the ingredients to the recipe. Between the two curly braces (`{}`) is the recipe. In this case, add `x` and `y`. The `return` function indicates what should be returned at the end of the function:

```
> mySum <- function(x,y){
     ## define the function mySum with arguments x and y
     ## the curly brace { begins the function definition or recipe
     resultSum <- x + y # sum the two arguments
     return(resultSum) # return the result
 }#end the function definition or recipe
> newResult <- mySum(x=4,y=5)
> newResult
[1] 9
```

A lot more can be done with functions in R, but extensive function writing in R is beyond the scope of this text.

4.8 Reading and Writing Data

Typically the enumeration (the "data") is stored external to the R environment and is read (or imported) into R, usually as a data frame (see Section 4.6.4). There are many ways to read and write information to and from R. Here, we will discuss one format that is commonly used: a CSV file (comma-separated values).

Microsoft Excel is usually used to create a CSV file. Open Excel and enter the enumeration for each variable down the rows of a column. Have the first row be the names of the variables. Each column is a separate variable, and each row represents a different sample unit. Figure 4.2 is an example of entering an enumeration. There are two variables: grade point average (GPA) and major. There are four sample units (students), thus the sample size is 4. The two variables were measured on each student.

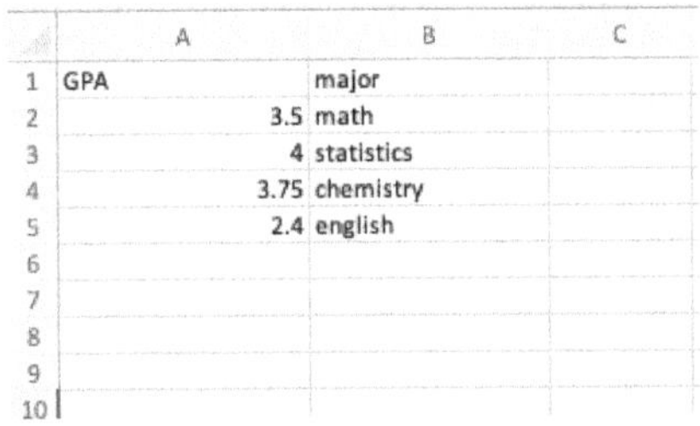

	A	B	C
1	GPA	major	
2	3.5	math	
3	4	statistics	
4	3.75	chemistry	
5	2.4	english	
6			
7			
8			
9			
10			

Figure 4.2: Example enumeration entered into Excel.

After the enumeration has been typed into Excel, go to the File menu and select Save As. Change the file format to be a CSV file. The name and the location of the file can also be changed. Save the CSV file on your computer.

When R is started it points (or looks) to a particular directory (or file folder). Usually this location is not somewhere helpful. The directory that R is pointing to needs to be changed before the CSV file can be read into R. To do this, select the File menu and the "change dir" or "change directory" option. Using the navigation window, select the location of the saved CSV file.

The example R code assumes that R is pointing to the directory that contains a file called `majorGPA.csv`. To read that file into R as a data frame and assign it to the variable called `df`, do the following:

```
> df <- read.csv("majorGPA.csv")
> df
   GPA      major
1 3.50       math
2 4.00 statistics
3 3.75  chemistry
```

```
4 2.40     english
```

For a data frame in R, the function `write.csv()` will create a CSV file in the directory in which R currently points:

```
> data(iris) # load a dataframe that is already in R
> ## the first argument is the data frame the second is the file name
> ## do not forget the .csv on the file name
> write.csv(iris, file='iris.csv')
```

This chapter provided a brief introduction to R. The interested reader can find lots of books and internet material to provide more information. The majority of the remaining chapters include a section showing how to do the math portion of the statistics within R.

Chapter 5

Summaries and Summary Tables

This chapter begins the process of using the enumeration. Looking at the whole enumeration is often confusing, and it is hard to determine an answer to the question of interest. If the enumeration is like a whole book, the summaries are like the CliffsNotes, the key ideas from the enumeration.

5.1 Terms

Enumeration: The complete listing of the outcomes from the variable for the entire sample.

Summary statistic: Information obtained from the sample that condenses the information from the enumeration into a few values or a single value.

Frequency (f): The count of how often a particular outcome occurs within the enumeration.

Relative frequency (rf): The proportion of how often a particular outcome occurs within the enumeration. Mathematically shown as rf =f/n (relative frequency = frequency/sample size). Relative frequency is a scaling of frequency between 0 and 1.

Percentage (%): The changing of the relative frequency scale to a 0 to 100 scale. Percent = 100*rf. Relative frequencies are hard to interpret, so we almost always interpret a rf as a percentage.

Table: A method of presenting summary statistics. Another common method of presenting summary statistics is graphically (Chapter 6).

An enumeration is created after a sample has been collected and every member of the sample has been measured using the variable. This enumeration is used to answer the question of interest; however, this may be difficult if the sample size (n) is large. To overcome this issue the enumeration is often condensed into summaries, which are typically presented in a table. The summaries statistics (frequency, relative frequency, and percentage) are presented in relationship to the outcomes of the variable.

5.2 Examples

1. What are U.S. college students' favorite online news source? A fictitious randomly selected sample of 50 U.S. college students were selected and asked, "What is your favorite online news source: theblaze.com, cnn.com, msnbc.msn.com, fox.com, or theonion.com?"

   Population: All U.S. college students at the time of the study.

   Sample frame: The complete listing of all U.S. college students.

   Sample unit: Any one U.S. college student.

   Sampling method: Simple random sampling. The sample was chosen randomly, so it should be a good representation of the population.

   Sample size: $n = 50$.

   Note: For the sample to be truly representative of the population it needs to be random and "large enough." What is large enough? This is hard to say, and it is a different size for each situation.

   Variable: Favorite online news source.

   Outcomes: theblaze.com (blaze), cnn.com (cnn), msnbc.com (msnbc), fox.com (fox), theonion.com (onion). For ease of talking and writing it's typical to give the outcomes shorter names, given in the parentheses. Note, the variable does not involve the sample or the population.

   Level of measurement: Categorical. The outcomes cannot be ordered in any objective way. The word "favorite" does not imply the level of measurement is ordinal. This word provides direction to the subjects on how to chose an outcome.

   Enumeration: The results of the sample:

onion	onion	blaze	cnn	onion	blaze	blaze	onion	blaze	cnn
blaze	fox	onion	blaze	cnn	fox	msnbc	msnbc	fox	cnn
blaze	onion	cnn	msnbc	blaze	fox	blaze	onion	blaze	cnn
cnn	cnn	fox	blaze	msnbc	cnn	fox	fox	blaze	blaze
cnn	cnn	blaze	blaze	fox	blaze	blaze	fox	blaze	blaze

The complete listing of the information from the sample is not very helpful. It is hard to answer a few simple questions such as:

(a) How many times is Blaze listed? CNN? MSNBC? Fox? Onion?

(b) Which online news source is listed the most? This is our primary question.

Summary statistics: The first logical thing to find is how many times each outcome occurs. This is called the **frequency** or **frequency count**. Thus, look at the enumeration and count how many times Blaze is listed, then do the same for the other outcomes. The results could be given like this:

Blaze occurs 19 times, or Blaze (f = 19)

CNN occurs 11 times, or Cnn (f = 11)

Fox occurs 9 times, or Fox (f = 9)

MSNBC occurs 4 times, or MSNBC (f = 4)

Onion occurs 7 times, or Onion (f = 7)

Note: What is the sum of all the frequencies? It will always sum to the sample size (n).

We could also summarize the results with relative frequencies. This is a scaling of the frequency to be between 0 and 1, $rf = f/n$:

Blaze rf $=19/50 = 0.38$

CNN rf $=11/50 = 0.22$

Fox rf $=9/50 = 0.18$

MSNBC rf $=4/50 = 0.08$

Onion rf $=7/50 = 0.14$

Note: The sum of all the rf values equals 1: $0.38 + 0.22 + 0.18 + 0.08 + 0.14 = 1$

In theory, the sum of all the rf always equals 1. In practice, with rounding errors it might not be exactly 1, but close.

Finally, the results can be summarized by percents: % = 100*rf:

Blaze percent = 100*0.38 = 38%

CNN percent = 100*0.22 = 22%

Fox percent = 100*0.18 = 18%

MSNBC percent = 100*0.08 = 8%

Onion percent = 100*0.14 = 14%

Note: The sum of all the percents equals 100%: 38% + 22% + 18% + 8% + 14% = 100%

The same rounding issues as with rf can happen with percents.

Table summarization:: Here, we take the previous summarization and put it into a table:

News Source	f
Blaze	19
CNN	11
Fox	9
MSNBC	4
Onion	7
Total	50

Table 5.1: Frequency table for the preferred online news source.

Table 5.1 presents the frequency table. It is easy to answer our two initially posed questions:

(a) How frequently does each online news source appear in our sample?

Answer: Blaze (19), CNN (11), Fox (9), MSNBC (4), Onion (7)

(b) Which online news source is listed most frequently?

Answer: Blaze: 19 of the college students in the sample consider theblaze.com their favorite online news source.

For each of the frequencies, we cannot tell if they are large or small without comparing a frequency to the rest of the frequencies. This is why we create the relative frequency (Table 5.2) or percentage (Table 5.3) tables. By bringing in the information of the sample size (n), the summary information is immediately more meaningful. For instance, CNN occurs 22% of the time in our sample, so it gives a sense of how often we should expect this result in the population as well. Relative frequencies are used for calculations. Percentages are used to interpret the relative frequencies.

News Source	rf
Blaze	0.38
CNN	0.22
Fox	0.18
MSNBC	0.08
Onion	0.14
Total	1

Table 5.2: Relative frequency table for the preferred online news source.

News Source	%
Blaze	38
CNN	22
Fox	18
MSNBC	8
Onion	14
Total	100

Table 5.3: Percent table for the preferred online news source.

Now consider a few questions:

(a) Interpret the frequency for the Onion.

Answer: Seven of the 100 college students favorite online news source is the Onion.

(b) How many college students' favorite online news source is CNN, Fox, or MSNBC?

Answer: $11 + 9 + 4 = 24$

(c) What is the rf for the Blaze?

Answer: rf $= 0.38$

(d) Interpret the rf for the Blaze.

Answer: 38% of college students favor the Blaze as their online news source.

(e) What is the rf of college students who did not choose MSNBC?

Answer: $0.38 + 0.22 + 0.18 + 0.14 = 0.92$ or $1 - 0.08 = 0.92$

It is either the sum of the other rf values or 1 minus the rf for MSNBC.

(f) What is the rf of college students who did not choose the Blaze or CNN?

Answer: $1 - 0.38 - 0.22 = 0.4$ or $0.18 + 0.08 + 0.14 = 0.4$

(g) What is the rf of college students who like either Fox or the Onion?

Answer: $0.18 + 0.14 = 0.32$

(h) Interpret the rf of college students who like either Fox or the Onion.

Answer: 32% of college students like either Fox or the Onion as their online news source.

2. There has been a long-standing debate between which brand of truck is better: Chevy or Ford. To answer this question, a fictitious random sample of ranchers and farmers who live in Wyoming was collected. Each farmer/rancher was asked, "What is your primary work truck: Chevy (C), Ford (F), or Dodge (D)?" Here is the enumeration:

F F F F C D F F C C D C F F C C C D D C F F F
C F F D F F C C C C C C F F D F C F F F D D D
F F C F C F F F F F F C D C F F F F F C F F F
F F C F F C C F D C F C F F C F C F C D C D F
F F F F D D F F F C C C D F F F C F F D F F D
C C F D F C C C C F F C C F D C F D D C C F F
C D C F F F F F D F C F

It is always good to identify the parts and pieces of the scenario before doing any of the "math."

Population: All Wyoming ranchers and farmers.

Sample frame: A complete listing of all ranchers and farmers living in Wyoming.

Sample unit: Any one Wyoming rancher or farmer.

Sampling method: Simple random sampling.

Sample size: $n = 150$; the total number from the enumeration.

Variable: What is your primary work truck?

Outcomes: Chevy (C), Ford (F), and Dodge (D)

Level of measurement: Categorical. There is not an objective way to order the outcomes.

(a) Create a table with the frequencies, relative frequencies, and the percertages.

Answer: It is common to have a single table with multiple summary statistics included:

Truck Brand	f	rf	%
C	48	0.32	32
D	24	0.16	16
F	78	0.52	52
Total	150	1.00	100

(b) Which brand of truck is used primarily by ranchers and farmers in Wyoming?

Answer: Ford (F). This brand has the highest f, rf, and percent.

(c) Interpret the value of 48.

Answer: Of the 150 Wyoming farmers and ranchers measured, 48 indicated their primary work truck is a Chevy.

(d) Interpret the value of 0.52.

Answer: Of the 150 Wyoming farmers and ranchers measured, 52% indicated their primary work truck is a Ford.

5.3 Examples With R

1. Return to the online news source example (page 34):

```
> ## create a vector in R
> newsSource <- c('onion', 'onion', 'blaze', 'cnn', 'onion', 'blaze', 'blaze', 'onion',
                'blaze', 'cnn', 'blaze', 'fox', 'onion', 'blaze', 'cnn', 'fox',
                'msnbc', 'msnbc', 'fox', 'cnn', 'blaze', 'onion', 'cnn', 'msnbc',
                'blaze', 'fox', 'blaze', 'onion', 'blaze', 'cnn', 'cnn', 'cnn', 'fox',
                'blaze', 'msnbc', 'cnn', 'fox', 'fox', 'blaze', 'blaze', 'cnn', 'cnn',
                'blaze', 'blaze', 'fox', 'blaze', 'blaze', 'fox', 'blaze', 'blaze')
> ## the table function to get the frequencies
> newsDF <- as.data.frame(table(newsSource))
> newsDF$rf <- newsDF$Freq/sum(newsDF$Freq)
> newsDF$percent <- newsDF$rf*100
> newsDF

  newsSource Freq   rf percent
1      blaze   19 0.38      38
2        cnn   11 0.22      22
3        fox    9 0.18      18
4      msnbc    4 0.08       8
5      onion    7 0.14      14
```

The data frame `newsDF` has the f, rf, and percent for each of the news source outcomes.

2. Return to the work truck example (page 38):

```
> primaryTruck <- c('F', 'F', 'F', 'F', 'C', 'D', 'F', 'F', 'C', 'C', 'D', 'C', 'F',
                   'F', 'C', 'C', 'C', 'D', 'D', 'C', 'F', 'F', 'F', 'C', 'F', 'F',
                   'D', 'F', 'F', 'C', 'C', 'C', 'C', 'C', 'C', 'F', 'F', 'D', 'F',
                   'C', 'F', 'F', 'F', 'D', 'D', 'D', 'F', 'F', 'C', 'F', 'C', 'F',
                   'F', 'F', 'F', 'F', 'F', 'C', 'D', 'C', 'F', 'F', 'F', 'F', 'F',
                   'C', 'F', 'F', 'F', 'F', 'F', 'C', 'F', 'F', 'C', 'C', 'F', 'D',
                   'C', 'F', 'C', 'F', 'F', 'C', 'F', 'C', 'F', 'C', 'D', 'C', 'D',
                   'F', 'F', 'F', 'F', 'F', 'D', 'D', 'F', 'F', 'F', 'C', 'C', 'C',
                   'D', 'F', 'F', 'F', 'C', 'F', 'F', 'D', 'F', 'F', 'D', 'C', 'C',
                   'F', 'D', 'F', 'C', 'C', 'C', 'C', 'F', 'F', 'C', 'C', 'F', 'D',
                   'C', 'F', 'D', 'D', 'C', 'C', 'F', 'F', 'C', 'D', 'C', 'F', 'F',
                   'F', 'F', 'F', 'D', 'F', 'C', 'F')
> truckDF <- as.data.frame(table(primaryTruck))
> truckDF$rf <- truckDF$Freq/sum(truckDF$Freq)
> truckDF$percent <- truckDF$rf*100
> truckDF

  primaryTruck Freq   rf percent
1            C   48 0.32      32
2            D   24 0.16      16
3            F   78 0.52      52
```

Chapter 6

Graphs

Graphs are visual representations of table summaries. Actually, a graphical summary can only be constructed after a table summary is created. Why create one? A picture is worth a thousand words. If done correctly, a graphical summary can convey a lot of information easily and well. On the other hand, a bad or wrong graph is worth a thousand lies.

Recall from Section 2.5 that all variables are treated as either categorical or ratio. Graphs will be described for these two level of measurement types only. Variables of other levels of measurement should be treated as categorical or ratio, as outlined in Section 2.5.

6.1 Terms

Horizontal axis: The axis on which the variable outcomes are located.

Vertical axis: The axis where the summaries (f, rf, %) of the outcomes are displayed.

Bar graph: A graph that displays a **categorical** variable, with bars in a vertical context. This graph is best at comparing one outcome to another outcome. The bar of one outcome does not touch the bar of another outcome. A bar graph is also referred to as a bar chart.

Pie graph: A graph that displays a **categorical** variable's outcomes in a circle. This graph is best at comparing one outcome to the whole. A pie graph is also referred to as a pie chart.

Pictogram: A graph that displays a categorical variable's outcomes as pictures. It is like a bar graph, but the bars are pictures instead.

Histogram: A graph that displays a **ratio** variable's outcomes as bars in a vertical context. The difference between a histogram and a bar graph is whether the bars touch. In a bar graph, the bars do *not* touch. In a histogram, the bars touch.

Line graph: A graph that displays a **ratio** variable's outcomes as a horizontal/vertical line.

6.2 Categorical Graphs

As indicated in some of the terms' definitions, where axes are applicable the outcomes will be on the horizontal axis and the summary statistic will be on the vertical axis. We will come back to the online news source example (Chapter 5) to construct some example graphs. Table 6.1 contains the summary statistics. It is common to put the f, rf, and percent all in one table. From the graph types described there are three ways of graphing a categorical variable.

News Source	f	rf	%
Blaze	19	0.38	38
CNN	11	0.22	22
Fox	9	0.18	18
MSNBC	4	0.08	8
Onion	7	0.14	14
Total	50	1	100

Table 6.1: Summary statistics of students' favorite online news source.

- A **bar graph** is used for f, rf, and % summary statistics.
- A **pie graph** is used for rf and % summary statistics.
- A **pictogram** is used for f, rf, and % summary statistics.

6.2.1 Bar Graph

In a bar graph, each outcome of the categorical variable is represented by a bar. The summary statistics (f, rf, or %) determine the height of the bar. In the horizontal axis of the graph we present the outcomes of the variable. In the vertical axis of the graph we present the summary statistic (f, rf, or %).

For the online news outlet example, we will create a frequency bar graph. The outcomes will be placed on the horizontal axis. The bar height for each outcome will be as tall as its outcome's frequency.

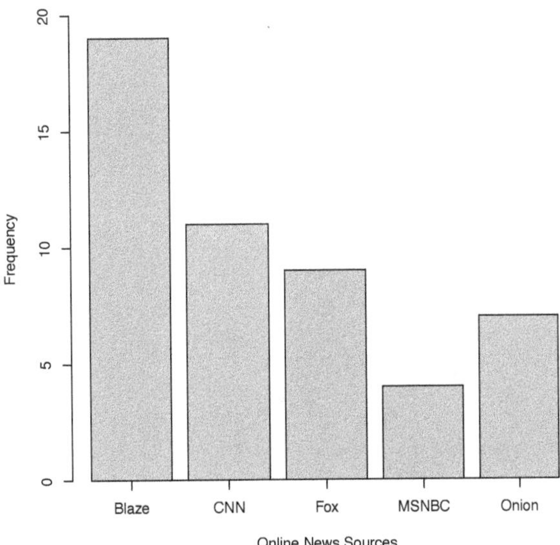

Figure 6.1: Frequency bar graph of students' favorite online news source.

It is easy to see from Figure 6.1 that the Blaze was rated the most favorable among college students. It is also easy to see that Fox is over twice as favorable as MSNBC. In general, it is easy to compare one outcome to another outcome.

6.2.2 Pie Graph

In a pie graph a circle is used to display the relative frequencies or percentages of the outcomes of a categorical variable through slices of the pie. Each pie slice in degrees $= 360^\circ * rf$. The degrees for each "slice" in the table are obtained from the rf, as presented in Table 6.2.

News Source	rf	Pie Slices in Degrees
Blaze	0.38	$(360)(0.38) = 136.8^\circ$
CNN	0.22	$(360)(0.22) = 79.2^\circ$
Fox	0.18	$(360)(0.18) = 64.8^\circ$
MSNBC	0.08	$(360)(0.08) = 28.8^\circ$
Onion	0.14	$(360)(0.14) = 50.4^\circ$
Total	1	360°

Table 6.2: Favorite online news sources' relative frequencies converted to degree angles.

There are many great ways of drawing the slices of the pie. The one I like is starting at the 12:00 position of the pie, then slicing off sections clockwise. This produces the pie graph presented in Figure 6.2. From the pie graph we can see that The Blaze and CNN take up almost three quarters of the pie. Also, MSNBC and

The Onion take up less than a quarter of the pie. This graph essentially compares individual or collections of outcomes to the whole.

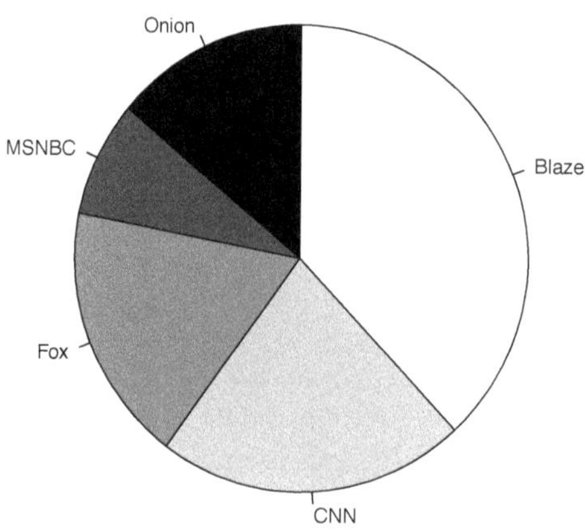

Figure 6.2: Relative frequency pie graph of students' favorite online news source.

6.2.3 Pictogram

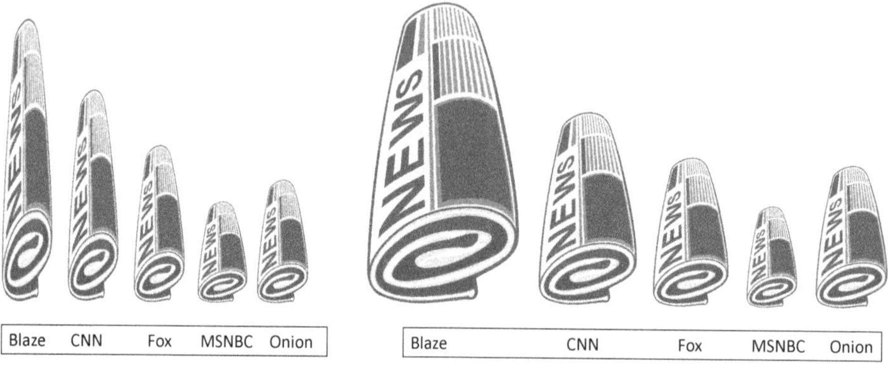

Figure 6.3: Pictogram of students' favorite online news source. The left figure changes only the height of the picture. The right figure scales the picture.

The professional and useful way of presenting summary statistics is the bar graph and occasionally a pie graph. The pictogram, on the other hand, is the consumer-friendly way of displaying information, mostly used by the news media. A pictogram is essentially a bar graph that replaces the bars with some representative figure/picture. Figure 6.3 presents an example using a picture of a newspaper. There are two basic ways for creating a pictogram. The first is to increase the height of each figure/picture to the height corresponding to the summary statistic (f, rf, or %). Because only the height changes, the figure/picture looks distorted. This is shown in the figure on the left. The second is to increase the whole figure/picture (height and width) for each outcome to represent the summary statistic. This is shown in the figure on the right. The bottom line with pictograms is that they are hard to create and can be misleading. It is good to know that pictograms exist, but this text will focus on the graphs that are used in professional journals.

6.3 Ratio Graphs

Histograms and **line graphs** are two ways to graph a single ratio variable. Both can be used with f, rf, or % summary statistics. For these graphs we cannot use the online news source example, as it is not a ratio variable. We need a ratio variable to create these graphs. For example, how much do full professors make at the local university? To answer this question suppose, that 50 full professors were asked, "What is your annual salary?" The enumeration is given. The numbers are in thousands of dollars, for example 1 = $1,000:

126 105 98 111 121 132 115 113 116 107 108 117 113 124 107 116 110

114 98 100 103 108 106 98 114 119 104 105 93 100 104 119 99 123

96 105 95 113 95 122 114 120 94 99 123 124 82 117 105 82

6.3.1 Histogram

The histogram is very similar to the bar graph in appearance. The biggest difference is that the bars touch in a histogram, and the bars do not touch in a bar graph. In order to construct a frequency histogram we need the frequency table first.

There are 29 unique salary values. A frequency summary table would be a little better (smaller) than the enumeration. Since there are so many outcomes, there would be a lot of bars. This would not be very useful or helpful. To get around this, we group the outcomes into ranges. These groups are called bins. From the frequency table we know the smallest outcome is 82 and the largest is 132. Now that the salary scores are in bins, a frequency table can be created, as presented in Table 6.3.

To construct the histogram, it will be a bar graph, except the bars of the histogram will touch one another. The bars will touch at the lower and upper points of the bins. For the salary example, this will be at 90, 100, 110, 120, and 130. Really, in a technical sense bin 1's salary is greater than and equal to 80 and less than 90 ($80 \leq$ salary(Bin1) < 90).

Faculty salary	f
Bin 1 (80 to 89)	2
Bin 2 (90 to 99)	10
Bin 3 (100 to 109)	14
Bin 4 (110 to 119)	15
Bin 5 (120 to 129)	8
Bin 6 (130 to 139)	1
Total	50

Table 6.3: Frequency table of faculty annual salary.

The frequency histogram is presented in Figure 6.4. From the histogram we see that most faculty make between \$110k and \$120k each year. This is the peak, and few faculty make more or less than the peak. More faculty make less than \$110k per year than faculty who make more than \$120k per year. Most faculty make between \$90k and \$120k. These are just a few of the statements that can be made from this graph.

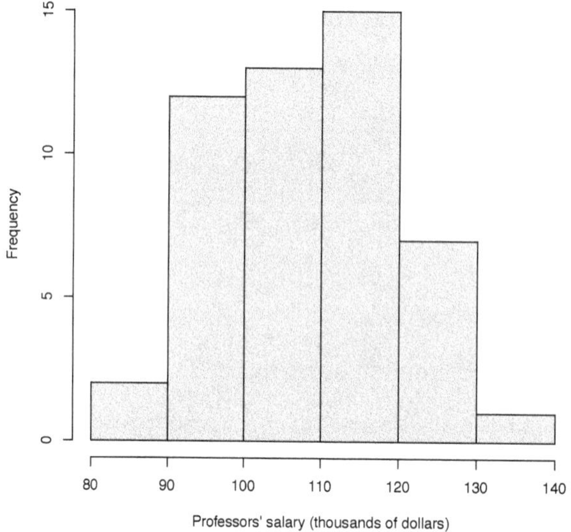

Figure 6.4: Frequency histogram of professors' annual salary.

6.3.2 Line Graph

A frequency table with bins is needed to create a line graph. Instead of bars there will be a single point. If we were to create a line graph over a histogram, the points would be on the top of each bar, halfway between the left and right edges of the bar. Then a line would connect the points. The midpoints are the point halfway between the right and left edges of each bar. Table 6.4 has the midpoints and calculations. We will construct a percentage line graph.

Bin	Midpoint	%
Bin 1 (80 to 89)	$(80 + 89)/2 = 84.5$	4
Bin 2 (90 to 99)	$(90 + 99)/2 = 94.5$	20
Bin 3 (100 to 109)	$(100 + 109)/2 = 104.5$	28
Bin 4 (110 to 119)	$(110 + 119)/2 = 114.5$	30
Bin 5 (120 to 129)	$(120 + 129)/2 = 124.5$	16
Bin 6 (130 to 139)	$(130 + 139)/2 = 134.5$	2

Table 6.4: Midpoint values for the bins from professors' salary with the percentage for each bin.

Now remember back to the good-old days in algebra class. Think of the midpoint and percent as the order pairs (X, Y), where X = midpoint and Y = percent. After the points are plotted, start at the left and connect the points with a straight line. Figure 6.5 presents the line graph.

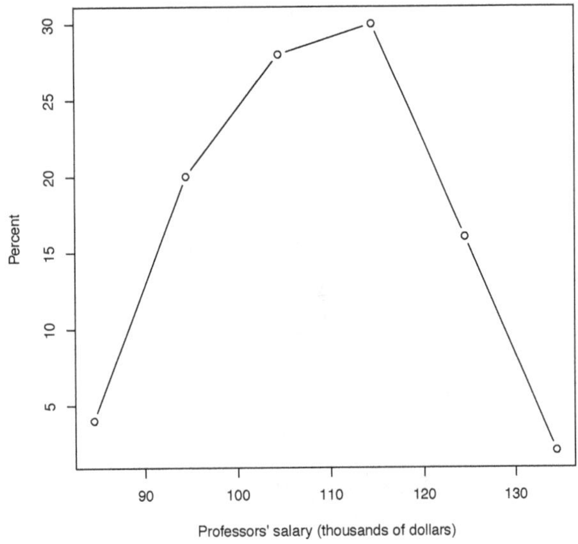

Figure 6.5: Percent line graph of professors' annual salary.

The line graph really tells the same story as the histogram, but the histogram is more common. Line graphs are most common in an area of statistics called time series: A variable is measured over time, and time is placed in the horizontal axis. Time series is beyond the scope of this text.

6.4 Problems in Graphs

The point of making a graph is to communicate summary statistics. Anything that makes the communication inaccurate or the results hard to understand is a problem with the graph. There are several minor issues and at least two major issues that can occur. The first is using the wrong type of graph, and the second is misleading axes.

Wrong graph choice: This is a twofold problem. The first is mismatching the graph and level of measurement. The most common occurance of this is mixing up a histogram and a bar graph. Histograms are for a ratio variable, and the bars touch. Bar graphs are for a categorical variable, and the bars do not touch. The second is choosing a graph that makes the results harder to understand or communicate. Figure 6.6 is an example of this problem. Without knowing anything else, we know the outcomes are the letters. The variable is "What is your favorite letter?" Even knowing that, it is hard to read this pie graph. We know "J" is the favorite letter. To know how the letter "S" compares to the whole is hard to tell. There are too many outcomes. When there are many outcomes, a bar graph should be used.

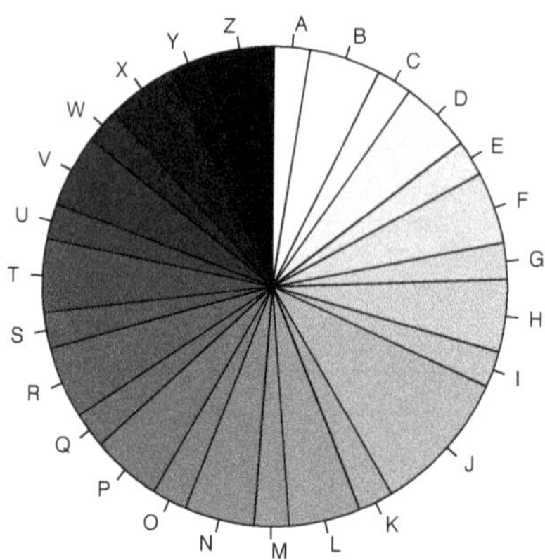

Figure 6.6: Pie graph with too many outcomes.

Axis problems: These can happen for bar graphs, histograms, or line graphs—basically any graph with axes. A bar graph will be used to illustrate the problem. Consider the bar graph presented in Figure

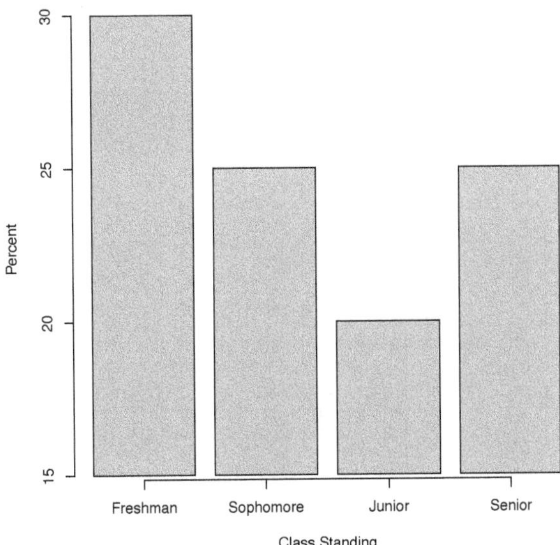

Figure 6.7: Percentage bar graph of class standing.

6.7 and see if you can see the problem. The graph shows percentages of students by class standing. At first glance nothing appears wrong. Let's use this bar graph to answer a few questions:

1. Most students have what class standing?

   Answer: Freshman. There is nothing wrong here; so far so good.

2. Are there twice as many sophomores as juniors?

   Answer: The sophomore bar is twice as big as the junior bar, but no, there are not twice as many sophomores as juniors. If you look at the vertical axis, it does not start at zero. Juniors make up 20% of the students, and 25% are sophomores. That is not double, but it appears that way based on the heights of the bars.

This graph also gives the impression that the percentages in each class standing are very different. The true graphs, as seen in Figure 6.8, gives a more accurate representation. We see from the correct graph that the percentages of class standings are not that different from one another.

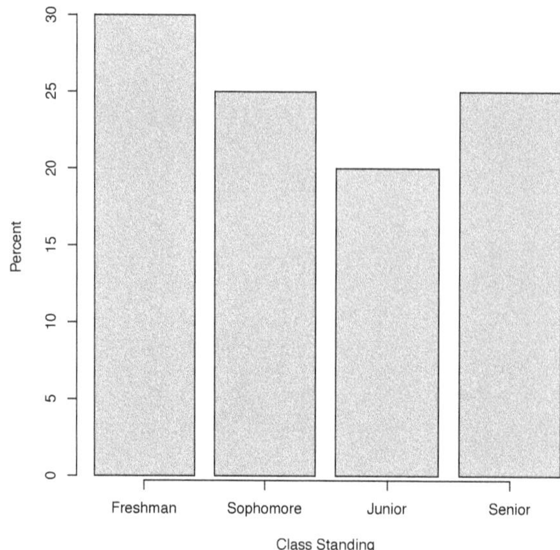

Figure 6.8: Correct percentage bar graph of class standing.

6.5 Examples

1. Return to the work truck example (page 38):

   (a) Which graph type is most appropriate?

   Answer: It depends. A histogram or a line graph is not appropriate because the variable is categorical, the answer is either a pie graph or bar graph. The bar graph is better at comparing one outcome to another. The pie graph is better at comparing one outcome to the whole. Thus, bar graph or pie graph are acceptable answers.

   (b) Which graph is the most appropriate in Figure 6.9?

   Answer: The most appropriate graph is **A**. Graph **B** is a histogram, which should be used for a ratio variable. Graph **C** is a line graph, which should be used for a ratio variable. The vertical axis for graph **D** does not start at zero. Even though graph **A** does not have labels for either axis, it is the most appropriate because it is a bar graph without any deception on either axis.

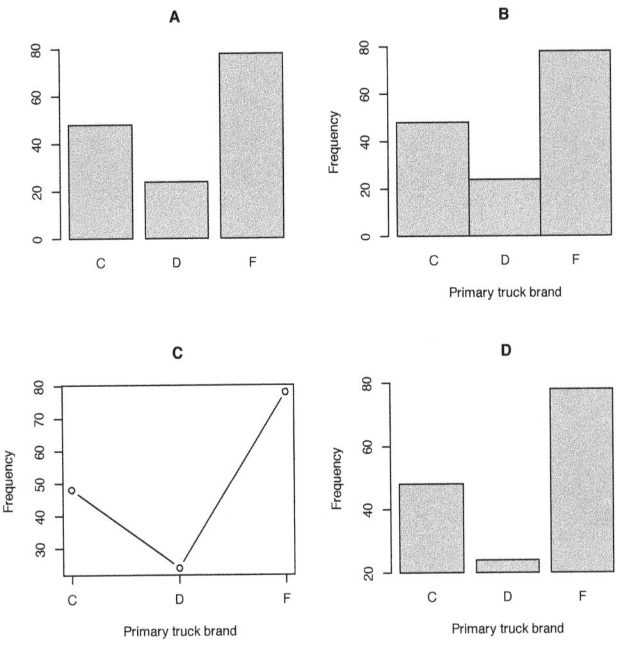

Figure 6.9: Graph choice from the primary truck example.

2. There is lots of money to be made or lost in the stock market. Suppose 20 fictitious companies were selected. The changes in their stock price during the last quarter were measured. The enumeration as follows:

-1.41	0.82	-0.23	1.58	0.79	0.83	-0.82	1.61	0.28	1.05
2.22	1.23	0.91	1.9	1.36	1.85	3.03	-0.61	0.75	0.61

(a) What is the level of measurement?

Answer: The variable is change in stock price. The level of measurement is ratio.

(b) Interpret the value of -1.41.

Answer: The stock price for that company decreased by $1.41.

(c) Use Figure 6.10 to answer the following questions:

i. Did the majority of companies have an increase or decrease in their stock price?

Answer: Increase. There are two bars below zero, which represents a frequency of four companies. Thus, the majority (or most of the companies) had an increase in their stock price.

ii. Did a majority of companies have an increase in their stock price greater than $2?

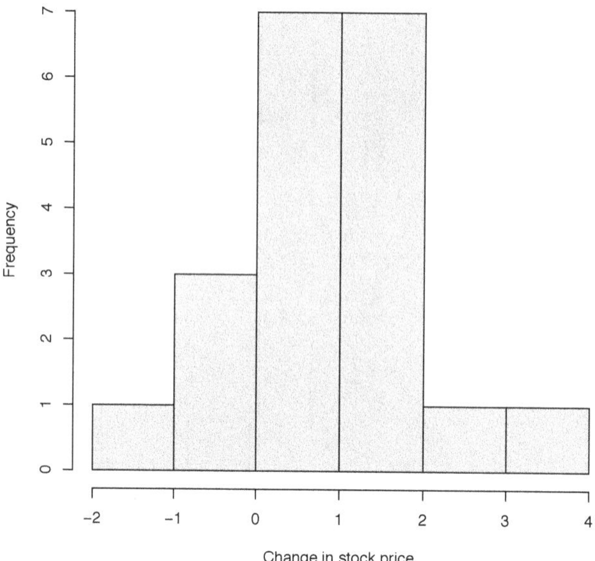

Figure 6.10: Histogram of the change in stock prices.

Answer: No. There are only two bars greater than \$2, which represented a frequency of two companies. Thus, the majority did not increase their stock price by \$2.

iii. Provide a range of values (two numbers) that describe the majority of companies' change in stock price.

Answer: Admittedly, this question is a little subjective. However, the range \$0 to \$2 seems reasonable. That range has two bars inside of it; those bars represent a frequency of 14 companies. In the sample, 14 companies represent a majority. The majority of companies within our sample had a stock increase between \$0 and \$2.

6.6 Example With `R`

A whole book could be devoted to graphics within `R`. The purpose here is to give the reader the very beginnings of `R` graphics. The `R` code is provided, but the actual graphics produced are omitted.

1. Return to the work truck example (page 38):

```
> ## named vector of the frequencies
> truckBrand <- c(C=48, D=24, F=78)
```

```
> ## frequency bar graph
> barplot(truckBrand)
> ## relative frequency bar graph
> barplot(truckBrand/sum(truckBrand))
> ## percent bar graph
> barplot(truckBrand/sum(truckBrand)*100)
> ## some additional arguments to make the plot nicer
> barplot(truckBrand,
        ylab='Frequency', # vertical axis label
        xlab='Truck Brand', # horizontal axis label
        ylim=c(0,80), # vertical axis bounds (min, max)
        col=c('red','green','blue'), # bar colors, one value means all bars the same
        )
> ## pie graph
> pie(truckBrand,
    clockwise=TRUE, # direction of the outcome, TRUE or FALSE
    col=c('red','green','blue') # slice colors
    )
```

2. Return to the stock price example (page 51):

```
> stockPrice <- c(-1.41, 0.82, -0.23, 1.58, 0.79, 0.83, -0.82, 1.61, 0.28, 1.05,
                  2.22, 1.23, 0.91, 1.9, 1.36, 1.85, 3.03, -0.61, 0.75, 0.61)
> ## histogram, with similar arguments as the barplot
> hist(stockPrice,
     xlab='Freq',
     ylab='Change in stock price',
     main='This is the title', # plot title
     col='blue'
     )
```

Credits

Chapter 7

Descriptors, Predictors, and Point Estimation

In Chapters 5 and 6 we discussed ways to summarize a enumeration in tables and graphs. This took an enumeration down to a smaller number of summary statistics. In this chapter we want to take it down to one value—a summary statistic. These single-value summaries will be used as descriptors and predictors:

Descriptor: Takes a single value used for describing the sample. There are several ways to describe a sample. The largest outcome, the smallest outcome, the most frequent outcome, and the typical outcome are the most common forms of describing a sample.

Predictor: Takes a single value and uses it to make predictions about a subject or on the population parameter.

Look back at the online news outlets example (Chapters 5 and 6). Recall from the frequency table or the frequency bar graph that theblaze.com was picked as the favorite by college students. This describes the sample. A prediction is the "theblaze.com is what we would predict is the favorite online news source for the next college student we meet." The prediction is outside of the sample.

Recall from Section 2.5 that all variables are treated as either categorical or ratio. Descriptors, predictors, and point estimates will be described for categorical and ratio variables. Variables of other levels of measurement should be treated as categorical or ratio, as outlined in Section 2.5.

7.1 Terms

Mode: The most frequent outcome in the enumeration. The outcome with the highest frequency. The mode is one of the outcomes and is not the frequency of the outcome.

Median: The middle value in the enumeration. Also the most typical value. This is a good descriptor of the sample and can be used to make predictions.

Mean: The average or typical value in the enumeration. This can describe the sample, but the median does a better job. The mean is a better predictor than the median.

> **Note**: The use of the word "value" instead of "outcome" in the definition of mean and median is not an accident. The mode has to be an outcome in the enumeration. The mean and median do not have to be observed outcomes in the enumeration but do need to be possible outcomes of the variable. The mode, median, and mean are all measures of centrality. That is, they get some sense of the center, typical, or most representative value from the sample.

Proportion: Another name for a relative frequency. The word "proportion" is used more in terms of prediction and estimation, whereas rf is used as a descriptor.

Fair probability: All outcomes are equally likely to occur.

Point estimate: A single guess used for making a prediction.

7.2 Mode

The mode is the most occurring outcome, a typical value of the sample. The mode is the best representation of the sample when the variable is categorical. This is most easily understood with an example. The mode does not have a special symbol.

7.2.1 Examples

1. Return to the online news outlet example (page 34). Here is the frequency table again. The outcome with the largest frequency is The Blaze. The mode is theblaze.com. Please note the mode is *not* 19. The mode is an outcome, not the frequency. If we use the mode to make a prediction, we would have predicted the next college student measured favors theblaze.com as their online news source.

Network	f
Blaze	19
CNN	11
Fox	9
MSNBC	4
Onion	7
Total	50

Table 7.1: Frequency table for the preferred online news source.

2. Suppose a random sample of 100 Wyoming residents was taken and each were asked, "What is your political affiliation: Republican, Democrat, or Independent?"

   Population: All Wyoming residents

   Variable: Someone's political affiliation

   Outcomes: Republican, Democrat, Independent

   Level of measurement: Categorical

   Here is the frequency summary table from the sample. The mode is Republican. The interpretation of the mode is this: Most people in the sample of Wyoming residents report their political affiliation to be Republican.

Affiliation	f
Republican	47
Democrat	34
Independent	19
Total	100

3. Suppose that six undergraduate college students were asked, "How many credits are you taking this semester?" Here is the enumeration:

   21 12 15 12 19 15

   Population: All undergraduate students

   Variable: Number of current credit hours

   Level of measurement: Ratio

   The frequency table of this enumeration will be the following:

   There are two outcomes with frequency 2. This enumeration has two modes: 12 and 15 credit hours. This is called bimodal. Most undergrads are taking either 12 or 15 credit hours this semester.

Credit Hours	f
12	2
15	2
19	1
21	1
Total	6

The mode is the only measure of centrality that can have two values. The median and the mean always have just one value.

7.3 Median

The median is the typical value. The median is the best descriptor of the sample. It is the middle value in the ordered enumeration. Because the enumeration needs to be ordered, a median *cannot* be calculated on a categorical variable. The median does not have a special symbol.

7.3.1 Examples

1. Suppose that nine randomly sampled elderly women in the United States were asked, "How often are you visited by friends or family: never, little, sometimes, often, always?" The enumeration is given:

   never sometimes never always sometimes always sometimes often often

   Population: All elderly women in the United States

   Variable: How often visited by friends or family

   Outcomes: never, little, sometimes, often, always

   Level of measurement: Ordinal treated as ratio

   The enumeration needs to be ordered first to find the medium:

   never never sometimes sometimes sometimes often often always always

   Now count to the middle from both sides:

never	never	sometimes	sometimes	sometimes	often	often	always	always
1	2	3	4		4	3	2	1

   The outcome "sometimes" is in the middle. Thus the median is "sometimes." The typical elderly woman in the United States is visited sometimes by friends and family.

2. Suppose that 20 college students were asked, "How many books did you buy for this semester?" The enumeration is as follows:

10   5   8   6   10   0   2   8   0   2   7   8   4   0   9   3   9   7   3   10

Population: All college students (A university was not indicated.)

Variable: Number of books bought this semester

Outcomes: 0, 1, 2, 3, . . . , etc. Lots of possible outcomes

Level of measurement: Ratio

Now put the enumeration in order:

0   0   0   2   2   3   3   4   5   6   7   7   8   8   8   9   9   10   10   10

There are the two middle numbers: 6 and 7:

$$\frac{6+7}{2} = \frac{13}{2} = 6.5$$

The median is 6.5. The typical student bought 6.5 books this semester. Wait, what? How can you buy a half a book? This is a case where the mathematics do not make practical sense. The typical student bought 6 to 7 books this semester.

3. Let's come back to the credit hour example. Here is the enumeration again:

   21   12   15   12   19   15

   Now we need to order the enumeration.

   12   12   15   15   19   21

   Note the sample size ($n = 6$) is an even number, so there are two middle numbers. When there are two middle numbers, we take the sum of the two numbers and then divide by 2:

$$\frac{15+15}{2} = \frac{30}{2} = 15$$

   Thus, the median is 15. The typical student this semester is taking 15 credit hours.

7.4 Mean

Everyone else calls the mean the average. Why is it called the mean in statistics? Probably to confuse Intro Stats students. The mean is the typical value in the sample. Recall the median is the best descriptor of the sample. The mean is the best predictor. Because of this, the mean gets a special symbol, μ. The Greek letter

μ (pronounced mu) represents the population mean. In practice, we never know this value. We estimate the population mean (μ) with the sample mean $\hat{\mu}$ (pronounced as mu hat).

Note: Anywhere a population parameter is given a symbol to represent it, that letter with a hat or carrot on top will be the estimate for the population parameter.

7.4.1 Formula

The formula for the mean is a little more complicated. The formula is presented in two ways: one using all mathematical notation, the other using words. This is done for completeness. Most other statistics texts will only use mathematical notation. A few symbols need to be defined before the formula can be understood:

x_i: The outcome value from the variable for the ith subject

$\sum$: The uppercase Greek letter sigma to sum all the elements

n: The sample size

$$\hat{\mu} = \frac{\sum_{i=1}^{n} x_i}{n} = \frac{x_1 + x_2 + \cdots + x_n}{n}$$

The nonmathematical notation formula is

$$\hat{\mu} = \frac{\text{sum of scores}}{\text{sample size}}$$

Summed the scores and divide by sample size.

7.4.2 Examples

1. Suppose seven college students were asked, "How many apps do you have on your smartphone?" The enumeration is given:

   143 183 130 180 128 79 171

   Population: All college students

   Variable: Number of apps on smartphone

   Level of measurement: Ratio

Now calculate the mean. The sample size is 7 ($n = 7$):

$$\hat{\mu} = \frac{143 + 183 + 130 + 180 + 128 + 79 + 171}{7} = \frac{1014}{7} = 144.8571$$

The mean is 144.86; usually round answers to two decimal places. Of these seven college students, the typical student has 145 apps on their smartphone (rounded to a whole number because you cannot have a partial app). This uses the mean as a descriptor of the sample. To use the mean as a predictor, I predict the typical college student has 145 apps on their smartphone, or I predict the next college student sampled will have 145 apps on their smartphone. The point estimate for the true mean number of apps on college students' smartphone is $\hat{\mu} = 144.86$.

2. More and more TV shows and networks are streaming on the internet. I asked 50 (fictitious) Americans aged 18–25, "In a typical week, how many hours of TV do you watch on the internet?" The enumeration is as follows:

13	15	12	1	7	8	16	17	7	17	8	25	32	9	8	0	40	5	33	8
15	14	37	17	11	35	19	15	6	34	24	26	35	0	7	15	25	28	11	2
11	12	30	19	13	29	2	34	1	22										

Population: All Americans aged 18–25

Variable: Number of hours watching internet TV

Level of measurement: Ratio

$$\hat{\mu} = \frac{\text{sum of scores}}{n} = \frac{830}{50} = 16.6$$

The mean is 16.6 hours. The typical American watches 16.6 hours of internet TV per week. For the next 18–25 year old American I meet, I expect they watch 16.6 hours of internet TV per week. The point estimate for the true mean hours of internet TV watched by 18–25 year old Americans is $\hat{\mu} = 16.6$ hours.

3. Suppose 30 United States voters were asked, "How likely are you to vote in the next election: 0 = not at all, 1 = maybe, 2 = possible, 3 = most likely, 4=definitely will?" The enumeration follows:

1	0	0	2	4	3	1	0	3	3	1	2	3	1	3
3	4	0	3	3	3	3	4	2	4	1	3	2	4	0

Population: All United States voters

Variable: Likelihood of voting in the next election

Level of measurement: Ordinal treated as ratio, because it has five outcomes

$$\hat{\mu} = \frac{\text{sum of scores}}{30} = \frac{66}{30} = 2.2$$

The mean is 2.2. The typical voter is 2.2 units likely to vote in the next election. This sounds a little weird; what does 2.2 mean on this scale? That is hard to say. We could round to the nearest whole number. The typical voter says it's possible they will vote in the next election. The point estimate of the true mean likelihood of voting in the next election is $\hat{\mu} = 2.2$.

7.5 Proportion

The letter P is given to represent the true population proportion. The symbol $\hat{P}$ is the point estimate for P. Since there are multiple outcomes, a subscript is used to denote what outcome is associated with what proportion. For example, if we have outcomes A, B, and C, then P_A is the true proportion for outcome A, and $\hat{P}_A$ is the estimated proportion or probability for outcome A. The same convention is done for each outcome. Proportions are used with categorical variables. When making a prediction, if the variable is ratio level of measurement, then use the mean; if categorical, typically use a proportion.

7.5.1 Fair Probability

The notion of fair probability comes up a lot, especially in games of chance, like flipping a coin or rolling dice. Fair probability is where every outcome has an equal chance of occurring. Most of the time (but not always), with a categorical variable we are interested in the existence, or lack thereof, of fair probability. The numerical value of the fair probability depends on the number of outcomes. The formula is

$$\tilde{P} = \frac{1}{\text{number of outcomes}}$$

Fair probability does not require an enumeration for the calculation, but collected information is used to assess if all the outcomes are equally likely. The fair probability symbol $\tilde{P}$ (pronounced p tilde) does not require a subscript. For example, a categorical variable with outcomes A, B, and C, assuming fair probability is

$$\tilde{P} = \tilde{P_A} = \tilde{P_B} = \tilde{P_C} = \frac{1}{\text{number of outcomes}} = \frac{1}{3}$$

Because all of the probabilities are the same, the fair probability is denoted without a subscript.

7.5.2 Examples

1. Suppose 100 U.S. voters aged 18–20 were asked, "In the next election are you going to vote Republican, Democrat, or Independent?" A summary of the enumeration is given:

Party	rf
Republican (R)	0.47
Democrat (D)	0.28
Independent (I)	0.25
Total	1

Summary statistic: The summary in the table is relative frequency

Population: All voters aged 18–20

All possible outcomes: Republican, Democrat, Independent

Level of measurement: Categorical

(a) What is the fair probability?

Answer:

$$\tilde{P} = \frac{1}{\text{number of outcomes}} = \frac{1}{3}$$

There are three outcomes.

(b) Interpret the fair probability.

Answer: I predict that 33% of 18–20 year old voters will vote Republican, 33% will vote Democrat, and 33% will vote Independent.

(c) What is the estimate of P_R?

Answer: The estimate is $\hat{P}_R = 0.47$.

(d) What is the interpretation of $\hat{P}_R$?

Answer: I predict that 47% of all 18–20 year old voters will vote Republican in the next election.

(e) What is the probability an 18–20 year old voter will vote Independent? Include the interpretation.

Answer: $\hat{P}_I = 0.25$; I predict 25% of voters 18–20 years old will vote Independent in the next election.

(f) What is the probability an 18–20 year old will not vote Democrat in the next election?

Answer: $\hat{P}_{notD} = .25 + .47 = .72$ or $\hat{P}_{notD} = 1 - .28 = .72$

The probability of not voting Democrat is either the sum of the other two or 1 minus the probability of voting Democrat.

2. Suppose 160 college students were asked, "Which type of grading system would you prefer: letter grade only or plus/minus letter grade?" Results:

System	rf
Letter only	0.65
Plus/minus	0.35
Total	1

Summary statistic: The summary in the table is relative frequency

Population: All college students

All possible outcomes: Letter only or plus/minus

Level of measurement: Categorical

(a) The administration assumes fair probability: What percent of students favors letter grades only system?

Answer: There are two outcomes: $\tilde{P} = 1/2$.

I predict that 50% of students favor the letter grade only system.

(b) What percent of the sample favors a letter grades only system?

Answer: 65% of the 160 measured students favor a letter grade only system.

(c) What percent of all college students favors a letter grades only system?

Answer: I predict 65% of students favor a letter grades only system.

Note: The answer to questions 2b and 2c are very similar. Question 2b and its answer use the proportion as a descriptor. Question 2c and its answer use the proportion as a predictor.

7.6 Examples With R

1. Return to the credit hour example (page 56). R does not have a function for the mode. However, there is a function to obtain a frequency table:

```
> creditHours <- c(21, 12, 15, 12, 19, 15)

> table(creditHours)

creditHours

12 15 19 21

 2  2  1  1

> mean(creditHours)

[1] 15.66667

> median(creditHours)

[1] 15
```

From the frequency table (function `table`), the modes are the same as previously described. The function **mean** and **median** provide the summary statistic, as their name implies.

2. Return to the hours of TV scenario (page 60):

```
> tvSize <- c(13, 15, 12, 1, 7, 8, 16, 17, 7, 17, 8, 25, 32, 9, 8, 0, 40, 5, 33,
              8, 15, 14, 37, 17, 11, 35, 19, 15, 6, 34, 24, 26, 35, 0, 7, 15, 25,
              28, 11, 2, 11, 12, 30, 19, 13, 29, 2, 34, 1, 22)

> mean(tv)

[1] 16.6

> median(tv)

[1] 15
```

3. Return to the voting scenario (page 62):

```
> vote <- c("R", "I", "I", "I", "I", "I", "R", "D", "R", "I", "I", "R", "R", "D",
            "I", "D", "I", "D", "R", "R", "R", "D", "I", "R", "R", "I", "R", "R",
            "D", "I", "R", "R", "D", "I", "R", "R", "R", "R", "D", "R", "R", "D",
            "D", "I", "I", "D", "D", "R", "I", "D", "R", "D", "R", "R", "R", "D",
            "D", "D", "D", "R", "D", "D", "R", "R", "R", "R", "R", "R", "I", "I",
            "R", "R", "R", "I", "I", "R", "R", "I", "D", "R", "D", "R", "R", "D",
```

```
        "D", "R", "D", "I", "R", "I", "R", "D", "R", "D", "D", "I", "R", "R",
        "I", "R")
> table(vote)
vote
 D  I  R
28 25 47
> voteDF <- as.data.frame(table(vote))
> voteDF$rf <- voteDF$Freq/sum(voteDF$Freq)
> voteDF$percent <- voteDF$rf*100
> voteDF
  vote Freq   rf percent
1    D   28 0.28      28
2    I   25 0.25      25
3    R   47 0.47      47
```

Chapter 8

Deviation and Variance

Now that we have used statistics (mean and proportion) to make predictions about the population, this leads to an important question: How good is the prediction? Assessing the quality is very important and can lead in two directions. One is expanding the point estimate to a range of estimates (Chapter 10). The other is doing hypothesis testing (Chapter 11). Before either of these can be done, the concept of variability needs to be discussed.

8.1 Terms

Deviation: How different a score is from the mean. Specifically, whether the mean underpredicts or overpredicts an individual score.

Standard deviation: The typical or "average" deviation. The square root of the variance.

Variance: A value of the variability of the sample. The standard deviation squared.

8.2 Variability

To assess how good a prediction is, we need to know how much variability is in the sample that was used to calculate the point estimate. Consider the following two samples from the sample population:

Sample A:	4	5	5	6	5	5
Sample B:	0	0	0	0	10	20

The mean from sample A is

$$\hat{\mu}_A = \frac{4+5+5+6+5+5}{6} = \frac{30}{6} = 5$$

The mean from sample B is

$$\hat{\mu}_B = \frac{0+0+0+0+10+20}{6} = \frac{30}{6} = 5$$

For both samples the best prediction is $\hat{\mu} = 5$. But for sample A, 5 is a better predictor because most of the scores are 5 or really close to that. For sample B, 5 is not as good a predictor because there are no values close to 5. This leads to mistakes or deviations.

8.2.1 Deviation

The score minus the mean is the deviation. The deviation is a way of accessing the prediction. A deviation close to zero indicates the prediction is doing well. A deviation far from zero (either positive or negative) indicates the prediction is not doing well. The mean is the prediction, and the score value is the target the mean is trying to predict. The best way to see this is through an example. Let's come back to sample A. Notice how most of the deviations are zero. This indicates the prediction is near perfect. The deviation of -1 indicates the mean overpredicted by one unit. The deviation of 1 indicated the mean underpredicted by one unit.

Score	$(x - \hat{\mu})$
4	4 - 5 = -1
5	5 - 5 = 0
5	5 - 5 = 0
6	6 - 5 = 1
5	5 - 5 = 0
5	5 - 5 = 0

Remember, the mean is a single number that is the best predictor. From sample A, we have a sample size of $n = 6$. We also have six deviations. We want a single number to summarize the deviations. This summary of the deviation is called the standard deviation. The first logical choice to summarize the deviation is to take the mean, since the mean is the best predictor. The sum of the deviation divided by n:

$$\text{sum of deviation} = \sum_{i=1}^{6}(x_i - \hat{\mu}) = -1+0+0+1+0+0 = 0$$

Since the sum of the deviation equals zero, the mean of the deviation equals zero as well.

Note: **In any sample, the sum of the deviations is always equal to zero.**

Since the sum of the deviations is equal to zero, the mean of the deviations will not work. There are a few different strategies to get around this and get a summarized value of the deviations. We will not explore all of the possibilities and will square each of the deviations, which has nice theoretical properties and leads to the variance and standard deviation.[1]

8.2.2 Variance and Standard Deviation

The standard deviation is the quantity we want, but to get the standard deviation we first need the variance. The variance of the sample is given a special symbol: S^2. The formula for the variance is

$$S^2 = \frac{\sum_{i=1}^{n}(x_i - \hat{\mu})^2}{n-1}$$

To make sense of the variance formula we will use sample A again. The first column is the enumeration of

	Score	$(x - \hat{\mu})$		$(x - \hat{\mu})^2$	
	4	4 - 5 =	-1	(-1)(-1) =	1
	5	5 - 5 =	0	(0)(0) =	0
	5	5 - 5 =	0	(0)(0) =	0
	6	6 - 5 =	1	(1)(1) =	1
	5	5 - 5 =	0	(0)(0) =	0
	5	5 - 5 =	0	(0)(0) =	0
Sum:	30		0		2

the scores. The second column is the deviations, and the third is the squared deviations. From sample A, the sum of the squared deviations is 2. The variance is

$$S_A^2 = \frac{\text{sum squared deviation}}{n-1} = \frac{\sum_{i=1}^{n}(x_i - \hat{\mu})^2}{n-1} = \frac{2}{6-1} = \frac{2}{5} = 0.4$$

Why divide by $n - 1$? The simple answer is it gives nice theoretical properties. Calculating the variance can be tricky, so practice is always a good thing. A template is set up to calculate the variance for sample B. How can we put zero for the sum of the deviation before this is calculated? Because the sum of the deviations is always zero. Here is the variance from sample B: $S_B^2 = 70$.

Now that we have the variance for both samples, how do we find the standard deviation? The standard deviation is the square root of the variance. The symbol for standard deviation is S. The formula is

$$S = \sqrt{S^2}$$

[1] Books on statistical inference or mathematical statistics explain the theoretical properties.

	Score	$(x - \hat{\mu})$	$(x - \hat{\mu})^2$
	0	0-5 =	
	0	0-5 =	
	0	0-5 =	
	0	0-5 =	
	10	10-5 =	
	20	20-5 =	
Sum:	30	0	

For sample A the standard deviation is

$$S_A = \sqrt{S_A^2} = \sqrt{0.4} = 0.63$$

For sample B the standard deviation is

$$S_B = \sqrt{S_B^2} = \sqrt{70} = 8.367$$

Which is the better predictor, $\hat{\mu}_A$ or $\hat{\mu}_B$?

Answer: $\hat{\mu}_A$ is the better predictor because $S_A < S_B$.

8.3 Examples

1. Return to the smartphone scenario (page 59). The enumeration again is

   143 183 130 180 128 79 171

   Population: All college students

   Variable: Number of apps on smartphone

   Level of measurement: Ratio

   The mean is $\hat{\mu} = 144.86$. The calculation for the variance is in the table. The sum of the deviations

Number of apps	Deviation	Squared deviation
143	-1.86	3.45
183	38.14	1454.88
130	-14.86	220.73
180	35.14	1235.02
128	-16.86	284.16
79	-65.86	4337.16
171	26.14	683.45
1014	0	8218.85

column is -0.02. Wait, the sum of the deviation is always equal to zero, right? The reason the deviations do not sum to 0 is due to rounding error. If all the numbers were not rounded the deviations would sum to zero. Now that the table is complete we can calculate the variance.

$$S^2 = \frac{3.45 + 1454.88 + 220.73 + 1235.02 + 284.16 + 4337.16 + 683.45}{7 - 1} = \frac{8218.85}{6} = 1369.81$$

After the variance we can calculate the standard deviation:

$$S = \sqrt{1369.81} = 37.01$$

Interpretation: When using the mean to predict the number of apps on a college student's smartphone, on average this prediction will result in an error of about 37 apps.

2. Return to the voting scenario (page 60). The enumeration again is

1 0 0 2 4 3 1 0 3 3 1 2 3 1 3

3 4 0 3 3 3 3 4 2 4 1 3 2 4 0

$$\hat{\mu} = \frac{\text{sum of scores}}{30} = \frac{66}{30} = 2.2$$

Remember, this variable is an ordinal level of measurement, but it has five or more outcomes, so it is treated as a ratio level of measurement. Table 8.1 presents the calculated deviations and squared deviations. The sum of the squared deviations is

$$\sum_{i=1}^{30}(X_i - \hat{\mu})^2 = 54.8$$

The variance is

$$S^2 = \frac{\sum_{i=1}^{30}(X_i - \hat{\mu})^2}{30 - 1} = \frac{54.8}{29} = 1.89$$

The standard deviation is

$$S = \sqrt{S^2} = \sqrt{1.89} = 1.37$$

Interpretation: When using the mean to predict the likelihood of voters voting, the typical person is different from the mean of 2.2 by 1.37 units.

X_i	$(X_i - \hat{\mu})$	$(X_i - \hat{\mu})^2$
1	-1.20	1.44
0	-2.20	4.84
0	-2.20	4.84
2	-0.20	0.04
4	1.80	3.24
3	0.80	0.64
1	-1.20	1.44
0	-2.20	4.84
3	0.80	0.64
3	0.80	0.64
1	-1.20	1.44
2	-0.20	0.04
3	0.80	0.64
1	-1.20	1.44
3	0.80	0.64
3	0.80	0.64
4	1.80	3.24
0	-2.20	4.84
3	0.80	0.64
3	0.80	0.64
3	0.80	0.64
3	0.80	0.64
4	1.80	3.24
2	-0.20	0.04
4	1.80	3.24
1	-1.20	1.44
3	0.80	0.64
2	-0.20	0.04
4	1.80	3.24
0	-2.20	4.84

Table 8.1: Deviation and squared deviation calculations.

8.4 Examples With R

The R code for creating the vector from the enumeration will not be presented. It was done as shown in previous chapters.

1. Return to the smartphone scenario (page 59):

```
> mean(phone) # mean

[1] 144.8571

> var(phone) # variance

[1] 1369.81
```

```
> sd(phone) # standard deviation
[1] 37.01094
```

2. Return to the voting scenario (page 60):

```
> mean(vote) # mean
[1] 2.2
> var(vote) # variance
[1] 1.889655
> sd(vote) # standard deviation
[1] 1.374647
```

Chapter 9

Probability Distributions

The process of answering the question of interest includes these steps: A sample is collected, a variable is measured, and that collected information is used to answer the question of interest. Because a sample is not a perfect representation of the population and measurement is not perfect, there is uncertainty in our answer. A probability distribution is used to quantify that uncertainty.

There are several probability distributions. In this text only three distributions will be discussed. Normal distribution (Section 9.2) will be used to give a basic understanding of how probability distributions work. The t (Section 9.3.1) and χ^2 (pronounced as chi-squared; Section 9.3.2) distributions are introduced because they are used for some hypothesis testing situations (Chapter 11).

A probability distribution is a mathematical function, such as $f(x) = x^2$. However, this mathematical function is not a probability distribution. The exact mathematical functions for the probability distributions are beyond the scope of this text. However, understanding that they are mathematical functions is important. Just as a mathematical function can be drawn as a curve in a graph, so too can a probability distribution. The horizontal axis is the values for the distribution or the inputs to the mathematical function, and the vertical axis is the values of the probability distribution or the height of the curve. Because not all mathematical functions are probability distributions, this means that probability distributions are mathematical functions with some particular properties:

- The height of the curve (or the value of $f(x)$) is always ≥ 0.
- The area between the curve and the horizontal axis for the whole distribution is always 1.0.
- Probability distributions are either continuous or discrete.

- A discrete probability distribution only has probabilities associated with whole numbers on the horizontal axis.
- A continuous probability distribution has probabilities associated with whole numbers and fractional numbers on the horizontal axis.

An excellent student will always ask three questions when a new probability distribution is introduced:

- Is the distribution continuous or discrete?
- What is the support of the distribution?
- What is the support of the parameters of the distribution?

These questions might not make sense now, but they should after reading this chapter.

9.1 Terms

Discrete: Only whole values, not partial or fractional values. A variable with any of the levels of measurement (Chapter 2) can be discrete.

Continuous: Whole and partial (or fractional) values. A variable with a ratio (or interval) level of measurement (Chapter 2) can be continuous.

Probability distribution: A mathematical curve (or function) in which the values of the curve are always positive and the area under the whole curve is equal to 1.0.

Distributional parameter: A value in the mathematic function that can be changed, which changes the location and/or shape of the distribution curve. A probability distribution can have more than one distributional parameter. The parameters are similar to the parameters discussed in Chapter 3 but are not exactly the same. In this text, it will be assumed the distributional parameters are known.

Distributional support: The range (smallest and largest values) that can be put into the distribution function. In a graphical sense, this is the smallest and largest values possible on the horizontal axis.

Distributional parameter support: The range (smallest and largest values) of possible values for the distributional parameters.

Probability statement: The notation used to indicate a probability from a distribution, for example $P(B < c)$. This is a generic statement that says what the probability is of being less than c using the B probability distribution. The statement can be changed to a "greater than" or the probability between two values. If the distribution is continuous, then $P(B < c) = P(B \leq c)$ is true. Whether it is "less than" (this also applies for "greater than") or "less than or equal to" does not matter for a continuous distribution. This is not the case for a discrete distribution.

9.2 Normal Probability Distribution

In nature it is very common to see symmetry with the outcomes of a variable. For example, take the height of male college students. Most of their heights are around the mean, and there is approximately the same amount of really tall male students as really short male students. Because of this, the normal distribution was developed.

The normal distribution has two parameters: the mean and the standard deviation. The interpretation of these parameters is the same as discussed for the mean (Chapter 7) and standard deviation (Chapter 8). Here are the answers to the three questions discussed at the beginning of this chapter:

- The normal probability distribution is continuous.
- The support of the distribution is $-\infty$ (negative infinity) to ∞ (positive infinity).
- The mean can be any value between $-\infty$ and ∞, and the standard deviation must be greater than zero.

A generic normal probability distribution is presented in Figure 9.1. The mean is in the center of the distribution, and that is the location of the peak of the distribution. The distribution is symmetric about the mean. If you draw a vertical line at the mean and fold the distribution in half on that line, the curves on either side will be the same. The mean provides the location of the normal distribution. A normal distribution with a mean of 75 is to the right on the number line of a normal distribution with a mean of 7. The standard deviation (sd) is given as the width of the line with arrows at each end. The sd determines the width of the normal distribution.

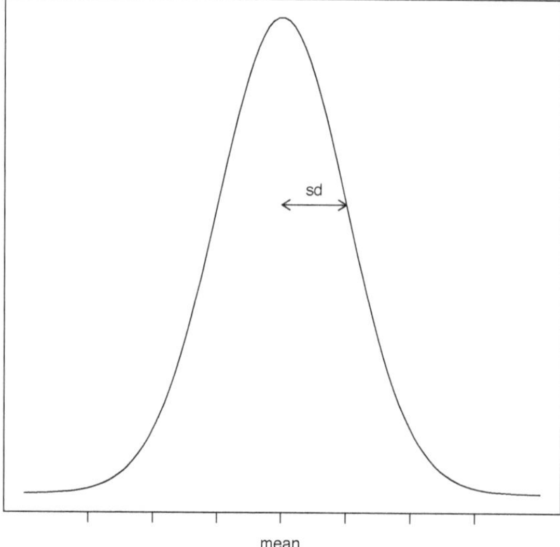

Figure 9.1: Generic normal probability distribution. The mean is at the center of the distribution, the location of the peak. The standard deviation (sd) is the width of the line with arrows.

9.2.1 Normal Distribution Empirical Rules

An area under the curve represents a probability. Because the sd corresponds to the width of the normal distribution there are some empirical rules associated with the normal distribution.

Let N represent the normal distribution; then probability statementslike $P(N < c)$ (What is the probability of being less than c using the normal distribution?) can be considered. Here are the normal distribution empirical rules:

- $P(N < mean) = P(N > mean) = 0.50$

  The probability of being less than (or greater than) the mean is 0.5.

- $P(mean - 1sd < N < mean + 1sd) = 0.68269$

  The probability of being within one standard deviation of the mean is 0.68269.

- $P(mean - 2sd < N < mean + 2sd) = 0.9545$

  The probability of being within two standard deviations of the mean is 0.9545.

- $P(mean - 3sd < N < mean + 3sd) = 0.9973$

  The probability of being within three standard deviations of the mean is 0.9973.

The empirical rules are visually displayed in Figure 9.2. The shaded area refers to the area of the curve that corresponds to the probability. In other words, the shaded area is the proportion of the curve that corresponds to that probability.

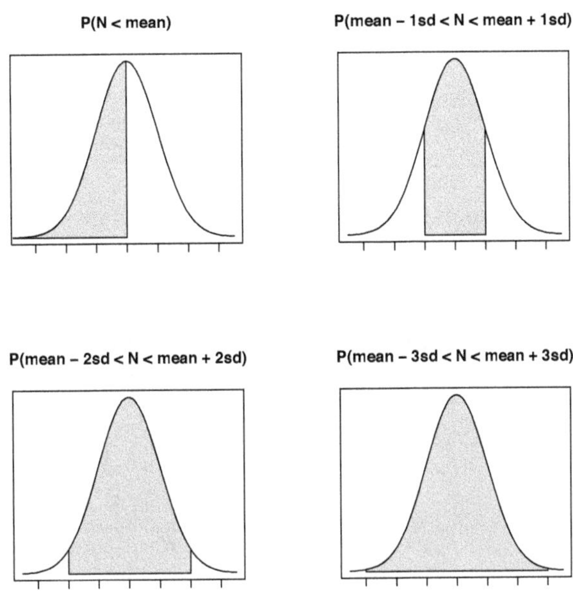

Figure 9.2: Empirical rules for the normal distribution. Shaded area corresponds to the probability statement above each panel.

9.2.2 Standard Normal Distribution

A normal distribution with a mean of 0 and a sd of 1 is a special case called the standard normal distribution. The standard normal distribution is given the letter Z as shorthand. A probability statement from any normal distribution can be transformed to a probability statement for the standard normal distribution. This is nice because only one table of probabilities is required to figure out any normal distribution probability statement. This is done with the following formula:

$$z = \frac{x - mean}{sd} \tag{9.1}$$

The probability statment $P(N < x) = P(Z < z)$ when x is converted to z using Eq. 9.1.

Sometimes we want to find the value of x for some given probability p, for example, find x such that

$P(N < x) = p$ is true. To do this, find z such that $P(Z < z) = p$ is true and then convert z to x using the following formula:

$$x = z(sd) + mean \tag{9.2}$$

9.2.3 Examples

1. Assume that male college students' height is normally distributed with a mean of 70 inches and a standard deviation of 2 inches.

   (a) What is the probability that a male student is shorter than 68 inches?

   First, write down what we know:

   - mean = 70
   - sd = 2
   - $x = 68$
   - We want "less than" because the question said "shorter."

   It is always good to draw a picture of what is required. Draw a normal curve with the peak at 70. Draw a vertical line at 68. Now shade in the left side of the vertical line because we want a less than probability, displayed in Figure 9.3. We expect a final probability less than 0.5 based on the figure.

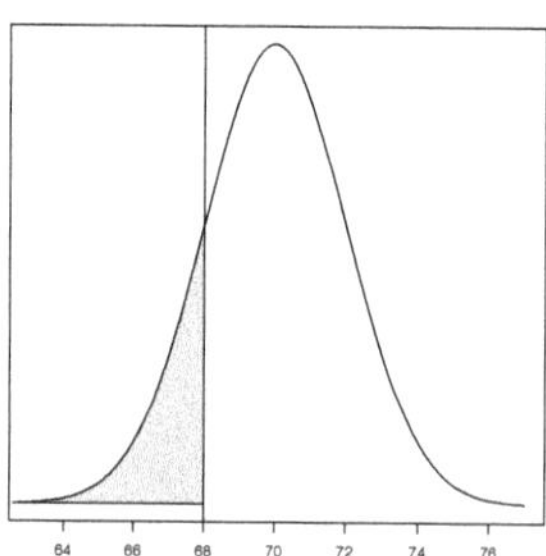

Figure 9.3: Normal distribution with mean = 70, sd = 2, and probability less than 68.

Now we can write a probability statement: $P(N < 68)$; convert this to a probability statement for the standard normal distribution.

$$z = \frac{x - mean}{sd} = \frac{68 - 70}{2} = -1$$

Now we know that

$$P(N < x) = P(Z < z)$$
$$P(N < 68) = P(Z < -1)$$

To find $P(Z < -1)$, we use one of the standard normal probabilities tables in Appendix B. We use Table B.1 because the z-score is negative. In the -1 row and the 0 column the value is 0.15866. Thus, $P(Z < -1) = 0.15866$, and therefore $P(N < 68) = P(Z < -1) = 0.15866$.

Answer: The probability that a male college student is shorter than 68 inches is 0.15866.

(b) What is the probability that a male college student is taller than 71 inches?

First, write down what we know:

- mean = 70
- sd = 2
- $x = 71$
- We want "greater than" because the question said "taller."

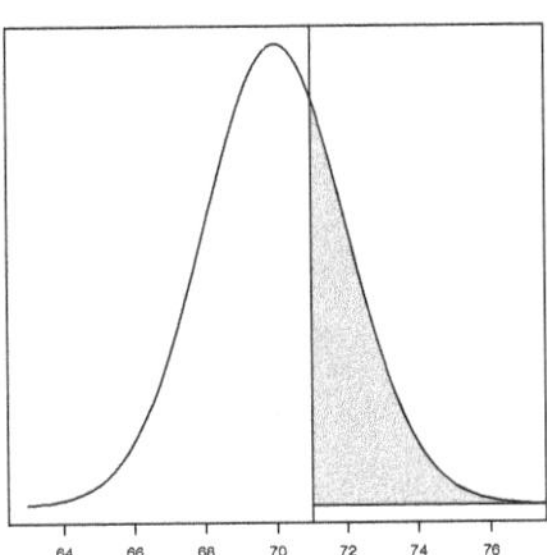

Figure 9.4: Normal distribution with mean = 70, sd = 2, and probability greater than 71.

It is always good to draw a picture of what is required. Draw a normal curve with the peak at 70. Draw a vertical line at 71. Now shade in the right side of the vertical line because we want a greater than probability. This is displayed in Figure 9.4. We expect a probability less than 0.5 based on the picture.

Now we can write a probability statement: $P(N > 71)$; convert this to a probability statement for the standard normal distribution:

$$z = \frac{x - mean}{sd} = \frac{71 - 70}{2} = 0.5$$

The tables in Appendix B only give less than probabilities. However, we know that the probability of everything (or the area under the whole curve) is 1.0. Thus,

$$\begin{aligned} P(Z > 0.5) &= 1 - P(Z < 0.5) \\ &= 1 - 0.69146 \text{ using the 0.5 row and the 0 column of Table B.2} \\ &= 0.30854 \end{aligned}$$

Answer: The probability that a male college student is taller than 71 inches is 0.30854.

(c) What is the probability that a male college student is between 67 and 72 inches?

First, write down what we know:

- mean = 70
- sd = 2
- $x_1 = 67$
- $x_2 = 67$

There are two x values because the question asks about the probability of being between two values. Figure 9.5 is the picture of the required probability: a normal distribution curve with the peak at 70. There is a vertical line at 67 and 72 with the area between them shaded.

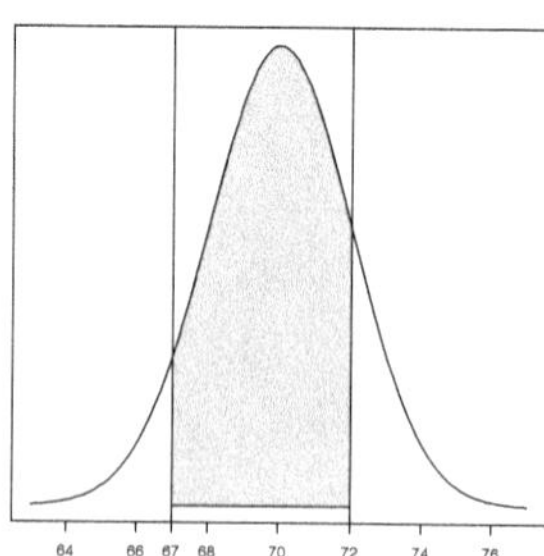

Figure 9.5: Normal distribution with mean = 70, sd = 2, and probability between 67 and 72.

The probability statement associated with Figure 9.5 and what is being asked in the question is $P(67 < N < 72)$. The Z tables in Appendix B do not provide probabilities for between two values. However, this can be split into two probability statements:

$$P(67 < N < 72) = P(N < 72) - P(N < 67) \tag{9.3}$$

The probability $P(N < 67)$ is the black shaded area in Figure 9.6, and the probability $P(N < 67)$ is the black shaded area plus the gray shaded areas in Figure 9.6. Thus, the difference is just the gray shaded area, which is the same as the shaded area in Figure 9.5.

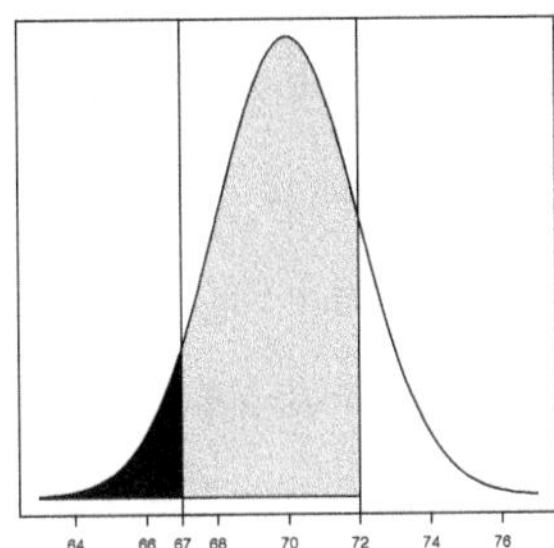

Figure 9.6: Normal distribution with mean = 70 and sd = 2. The black shaded area is $P(N < 67)$, and the black shaded area plus the gray shaded area is $P(N < 72)$. The gray shaded area is $P(67 < N < 72)$.

Now that the probability statement is broken down into probability statements (Eq. 9.3), we can convert to z-scores and look them up in one of the tables in Appendix B. Since there are two x values we need to calculate two z-scores:

$$z_1 = \frac{x_1 - mean}{sd} = \frac{67 - 70}{2} = -1.5 \tag{9.4}$$

$$z_2 = \frac{x_2 - mean}{sd} = \frac{72 - 70}{2} = 1 \tag{9.5}$$

Here is how this all comes together:

$$\begin{aligned}
P(67 < N < 72) &= P(N < 72) - P(N < 67) \text{ from Eq. 9.3} \\
&= P(Z < 1) - P(N < 67) \text{ from Eq. 9.5} \\
&= P(Z < 1) - P(Z < -1.5) \text{ from Eq. 9.4} \\
&= 0.84134 - 0.06681 \text{ using Tables B.2 and B.1} \\
&= 0.77453
\end{aligned}$$

Answer: The probability that a male college student is between 67 and 72 inches is 0.77453.

(d) What height corresponds to the 60th percentile? (Or, what is the 60th quantile? This is the same question worded differently.)

Write down what we know:

- mean = 70
- sd = 2
- probability = 0.6

Draw a picture of what we want to find. The picture in Figure 9.7 does that. A percentile (or quantile) is the same as a less than probability; thus, we will shade to the left of the vertical line. The percentile (or probability) is greater than 0.5; thus, the vertical line must be to the right of the mean, but we do not know exactly where. We are trying to find x such that the area to the left of x equals 0.6. This can be written as a probability statement: $P(N < x) = 0.6$.

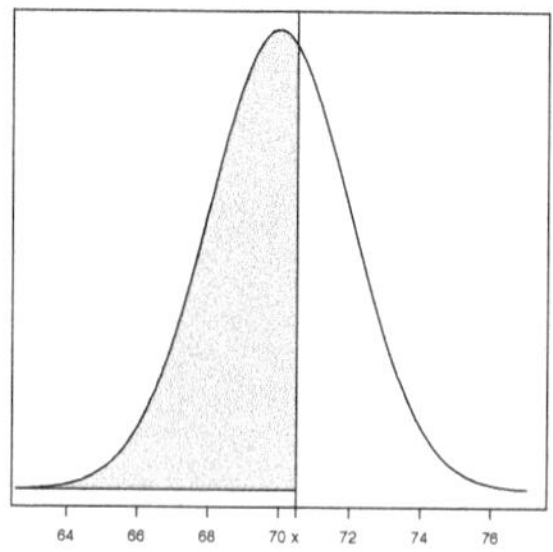

Figure 9.7: Normal distribution with mean = 70 and sd = 2. The shaded area is the probability of $0.6 = P(N < x)$, where x is unknown.

The tables in Appendix B can be used to find z such that $P(Z < z) = 0.6$ is true. For this we need to find the value 0.6 inside of Table B.2. The exact value will not be in the table, but we want to find the closest value that is above the value required. We find the following:

$$P(Z < 0.26) = 0.60257 \tag{9.6}$$

Therefore, the z-score is 0.26. This z-score can be converted back to x using Eq. 9.2:

$$\begin{aligned} x &= z(sd) + mean \\ &= 0.26(2) + 70 \\ &= 70.52 \end{aligned} \tag{9.7}$$

A z-score has be found from the probability, and that z-score has been converted back to x. Now put it all together:

$$\begin{aligned} 0.6 &= P(N < x) \\ &= P(Z < z) \\ &= P(Z < 0.26) \text{ using Eq. 9.6} \\ &= P(N < 70.52) \text{ using Eq. 9.7} \end{aligned}$$

Answer: A height of 70.52 inches corresponds to the 60th percentiles. In other words, a male college student with a height of 70.52 inches is predicted to be taller than 60% of male college students.

(e) What height corresponds to the top 10%?

What is known:

- mean = 70
- sd = 2
- probability = 0.1

The picture in Figure 9.8 represents what we are trying to find. The question asks for a "top percent," which means a "greater than" probability, so the area is to the right of the vertical line. Also it is a top percent that is smaller than 50%, so the vertical line must be above the mean. The value x is what is unknown and what we want to find. This can be written as a probability statement: $P(N > x) = 0.1$, where we want to find x.

The first thing to do is find the z-score, but the tables in Appendix B only give less than proba-

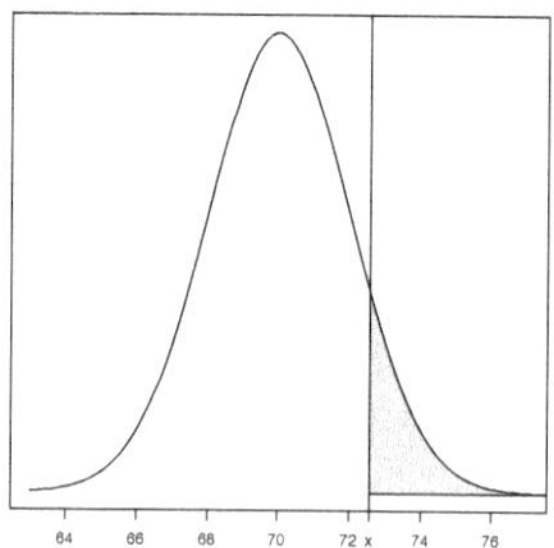

Figure 9.8: Normal distribution with mean = 70 and sd = 2. The shaded area is the probability of $0.1 = P(N > x)$, where x is unknown.

bilities. We need to convert to a less than probability first:

$$\begin{aligned} 0.1 &= P(N > x) \\ 1 - 0.1 &= 1 - P(N > x) \\ 0.9 &= 1 - P(N > x) \\ 0.9 &= P(N < x) \end{aligned} \tag{9.8}$$

Now Table B.2 can be used to find z such that

$$\begin{aligned} 0.9 &= P(Z < z) \\ 0.9 &= P(Z < 1.28) \end{aligned} \tag{9.9}$$

The value closest to 0.9, but below, was used to find the z-score. Now the z-score can be converted back to an x value:

$$x = z(sd) + mean = 1.28(2) + 70 = 72.56 \tag{9.10}$$

Put all of this together:

$$\begin{aligned}
0.1 &= P(N > x) \\
1 - 0.1 &= P(N < x) \text{ using Eq. 9.8} \\
1 - 0.1 &= P(Z < 1.28) \text{ using Eq. 9.9} \\
1 - 0.1 &= P(N < 72.56) \text{ using Eq. 9.10} \\
0.1 &= P(N > 72.56) \text{ convert back to greater than probability}
\end{aligned}$$

Answer: The top 10% is a height of 72.56 inches. In other words, a male college student with a height of 72.56 inches is predicted to be shorter than only 10% of male college students.

2. A certain professor teaching a class with 50-minute lectures does not always dismiss the class after exactly 50 minutes. Assume the time of class dismissal is normally distributed with a mean of 50 and a standard deviation of 4.

   For this example all of the questions will be asked first. The answers will be provided later; this is to provide an opportunity to practice before seeing the answers.

   (a) What is the top 20% for when class is dismissed?

   (b) What is the probability that the class is let out between 8 minutes early and 8 minutes late?

   (c) What is the probability the class is dismissed 2 minutes late or sooner?

   (d) What is the probability the class is dismissed 3 minutes early or later?

   (e) What is the 40th percentile (or quantile) for the time the class is dismissed?

   Work through the questions before reviewing the solutions.

   (a) What is the top 20% for when class is dismissed?

   What is known:

   - mean = 50
   - sd = 4
   - probability = 0.2

   Since we are looking for the top percentage, the probability statement is $P(N > x) = 0.2$. Figure 9.9 has the picture of what is required. The area is shaded to the right because the question asked for the top percentage. The vertical line at x is to the right of the mean because the top percentage is less than 50%. We want to find x, which leads to the probability statment: $P(N > x) = 0.2$.

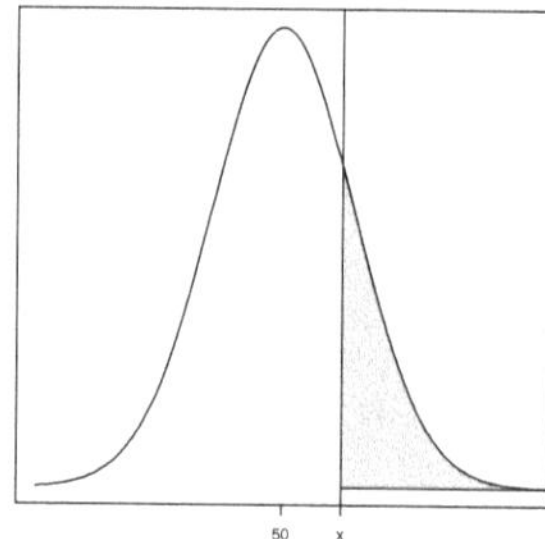

Figure 9.9: Normal distribution with mean = 50 and sd = 4. The shaded area is the probability of $0.4 = P(N > x)$, where x is unknown.

First convert the probability statement to a less than probability so that the z-score can be found from a table in Appendix B:

$$\begin{aligned}
0.2 &= P(N > x) \\
1 - 0.2 &= 1 - P(N > x) \\
0.8 &= 1 - P(N > x) \\
0.8 &= P(N < x) \\
0.8 &= P(Z < z) \\
0.8 &= P(Z < 0.84) \text{ using Table B.2}
\end{aligned} \tag{9.11}$$

The probability $P(Z < 0.84) = 0.79955$ is the closest to 0.8 without going over. Now convert the z-score back to an x value:

$$x = z(sd) + mean = 0.84(4) + 50 = 53.36 \tag{9.12}$$

Combine the results from Eq. 9.11 and 9.12:

$$\begin{aligned} 0.2 &= P(N > x) \\ 1 - 0.2 &= P(Z < 0.84) \text{ from Eq. 9.11} \\ 1 - 0.2 &= P(N < 53.36) \text{ from Eq. 9.12} \\ 0.2 &= P(N > 53.36) \text{ back to greater than probability} \end{aligned}$$

Answer: The top 20% of class dismissals is 53.36 minutes. In other words, only 20% of the time is the class dismissed later than 3.36 minutes late.

(b) What is the probability that the class is let out between 8 minutes early and 8 minutes late?

What is known:

- mean = 50
- sd = 4
- between 8 minutes early and 8 minutes late

If the class is dismissed 8 minutes early, then the class is dismissed at 42 minutes. If the class is dismissed 8 minutes late, then the class is dismissed at 58 minutes. We now also know the following:

- $x1 = 42$
- $x2 = 58$

Figure 9.10 represents what we want to calculate. The shaded area is the probability between 42 and 58 minutes. The probability statement to solve is $P(42 < N < 58)$.

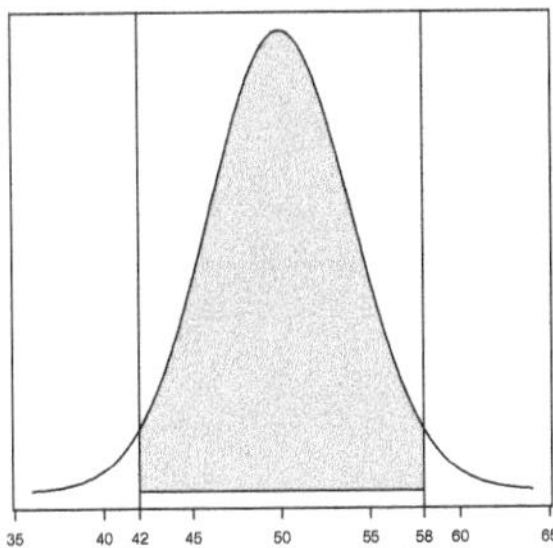

Figure 9.10: Normal distribution with mean = 50 and sd = 4. The shaded area is the probability of $P(42 < N < 58) = p$, where p is unknown.

First, convert the x values to z-scores:

$$z_1 = \frac{x_1 - mean}{sd} = \frac{42 - 50}{4} = -2 \tag{9.13}$$
$$z_2 = \frac{x_1 - mean}{sd} = \frac{58 - 50}{4} = 2 \tag{9.14}$$

Now find the probabilities and solve:

$$\begin{aligned} P(42 < N < 58) &= P(-2 < N < 2) \text{ using Eq. 9.13 and 9.13} \\ &= P(Z < 2) - P(Z < -2) \text{ probability for between two values} \\ &= 0.97725 - 0.02275 \text{ using tables in Appendix B} \\ &= 0.9545 \end{aligned}$$

Answer: The probability of being dismissed between 8 minutes early and 8 minutes late is 0.9545.

Or, the probability of being dismissed between 42 minutes and 58 minutes is 0.9545.

Note: There is a quicker way to answer this question:

$$mean - 2sd = 50 - 2(4) = 42$$
$$mean + 2sd = 50 + 2(4) = 58$$

Therefore,

$$\begin{aligned} P(42 < N < 58) &= P(mean - 2sd < N < mean + 2sd) \\ &= 0.9545 \text{ using the empirical rule (Section 9.2.1)} \end{aligned}$$

and the same answer is obtained.

(c) What is the probability the class is dismissed 2 minutes late or sooner?

What is known:

- mean = 50
- sd = 4
- $x = 52$ because that is 2 minutes late

Figure 9.11 represents the probability to be calculated. The area is shaded to the left because the question is the probability of that time or sooner. The probability statement is $P(N < 52)$.

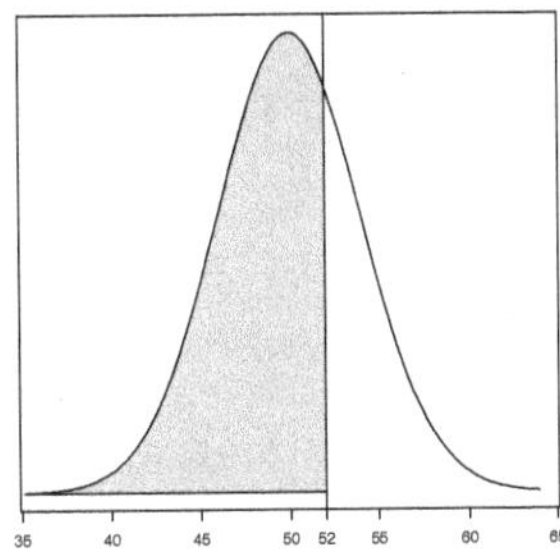

Figure 9.11: Normal distribution with mean = 50 and sd = 4. The shaded area is the probability of $P(N < 52) = p$, where p is unknown.

First, convert the x value to a z-score:

$$z = \frac{x - mean}{sd} = \frac{52 - 50}{4} = 0.5 \tag{9.15}$$

Now find the probability.

$$\begin{aligned} P(N < 52) &= P(Z < 0.5) \text{ using Eq. 9.15} \\ &= 0.69146 \text{ from Table B.2} \end{aligned}$$

Answer: The probability the class is dismissed 2 minutes late or sooner is 0.69146.

(d) What is the probability the class is dimissed 3 minutes early or later?

What is known:

- mean = 50
- sd = 4
- $x = 47$; that is 3 minutes early

Figure 9.12 represents the required probability. The area to the right is shaded because the question asks for the probability of a time or later. The probability statement is P(N > 47).

First, convert the x value to a z-score:

$$z = \frac{x - mean}{sd} = \frac{47 - 50}{4} = -0.75 \tag{9.16}$$

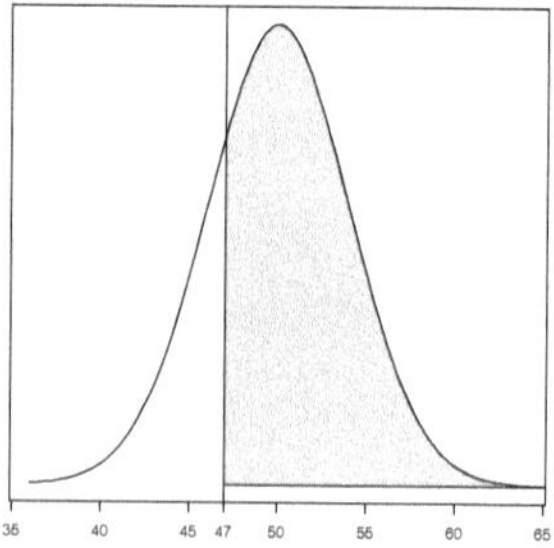

Figure 9.12: Normal distribution with mean = 50 and sd = 4. The shaded area is the probability of $P(N > 47) = p$, where p is unknown.

Now find the probability:

$$\begin{aligned} P(N > 52) &= P(Z > -0.75) \text{ using Eq. 9.16} \\ &= 1 - P(Z < -0.75) \text{ converts to less than probability} \\ &= 1 - 0.22663 \text{ from Table B.1} \\ &= 0.77337 \end{aligned}$$

Answer: The probability the class is dismissed 3 minutes early or later is 0.77337.

(e) What is the 40th percentile (or quantile) for the time the class is dismissed?

What is known:

- mean = 50
- sd = 4
- probability = 0.4

Figure 9.13 represents the probability and shows the x that is unknown. The area is shaded to the left because the problem gave a percentile (or quantile). The probability is less than 0.5, so the vertical line must be less than the mean. The probability statement is $P(N < x) = 0.4$, where x is what we want to find.

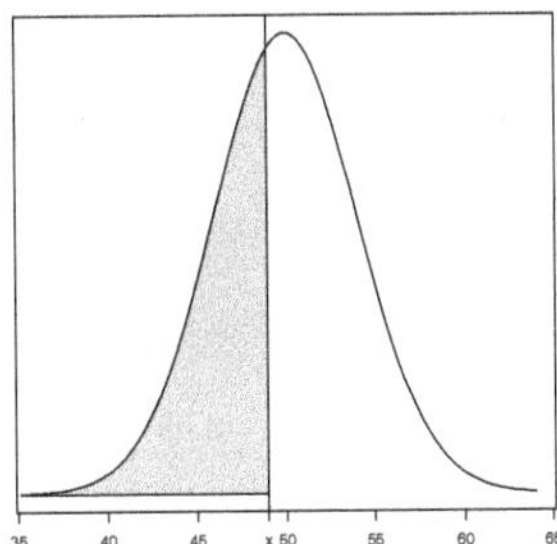

Figure 9.13: Normal distribution with mean = 50 and sd = 4. The shaded area is the probability of $P(N < x) = 0.4$, where x is unknown.

First, find the z-score associated with that probability:

$$\begin{aligned} 0.4 &= P(N < x) \\ &= P(Z < z) \\ &= P(Z < -0.26) \text{ from Table B.1} \end{aligned} \tag{9.17}$$

Now convert the z-score back to an x value:

$$\begin{aligned} x &= z(sd) + mean \\ &= -0.26(4) + 50 \text{ from Eq. 9.17} \\ &= 48.96 \end{aligned}$$

Answer: The 40th percentile (or quantile) for the time the class is dismissed is 48.96 minutes.

9.3 Other Distributions

There are lots of probability distributions. An interested reader should take a class on probability theory to learn more about probability distributions and learn more distributions. In this text two more distributions will be briefly covered: the t probability distribution and the χ^2 probability distribution. These distributions are used for some hypothesis testing (Chapter 11) situations. The nature of these distributions does not allow for a nice table to look up probabilities. Some probability-type questions can be determined from tables in Appendix A. It is good to have a basic understanding of t and χ^2 probability distributions.

9.3.1 t Probability Distribution

In this text, we will consider the t distribution with only a single parameter δ, called the degrees of freedom (δ is the Greek letter delta). Typically the distribution is denoted as t_δ, meaning a t distribution with δ degrees of freedom. The t_δ distribution is a continuous distribution. The support of the distribution is $-\infty$ to $+\infty$. The support of δ is greater than zero, and typically the degrees of freedom are integer values. A plot of t with different values of δ is presented in Figure 9.14. As seen, the distribution is symmetric about zero. As δ approaches infinity, the t distribution becomes the standard normal distribution.

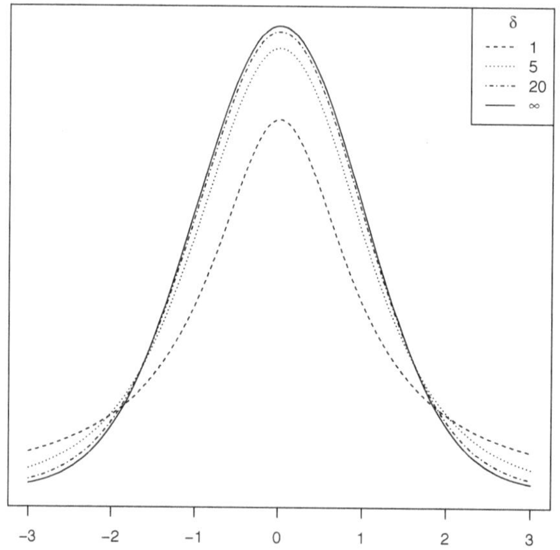

Figure 9.14: t distribution with several different values for δ, the degrees of freedom.

9.3.2 χ^2 Probability Distribution

In this text, we will consider the χ^2 distribution (pronounced as chi-squared distribution) with only a single parameter δ, called the degrees of freedom. Typically the distribution is denoted as χ^2_δ, meaning a χ^2 distribution with δ degrees of freedom. The χ^2_δ distribution is a continuous distribution. The support of the distribution is 0 to $+\infty$. The support of δ is greater than zero, and typically the degrees of freedom are integer values. A plot of the χ^2 distribution with different values of δ is presented in Figure 9.15. The peak of the distribution moves farther to the right as δ increases, but the peak is not over the value of δ.

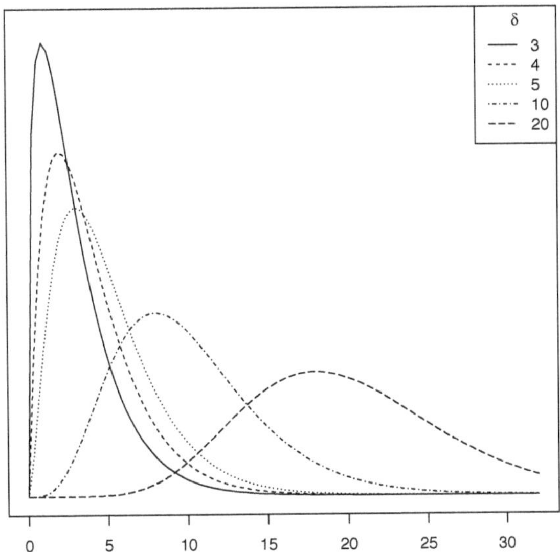

Figure 9.15: χ^2 distribution with several different values for δ, the degrees of freedom.

9.3.3 Examples

The column headings of Tables A.1 and A.2 are probability values. The first column of each table is the value of δ, or the degrees of freedom. The values on the inside of the table are the value for the top percent according to the column value.

1. What is the top 10% from the t distribution with $\delta = 7$?

   Answer: 1.415; use the 0.1 column and row 7 in Table A.2.

2. What is the top 10% from the χ^2 distribution with $\delta = 7$?

   Answer: 12.017; use the 0.1 column and row 7 in Table A.1.

3. What is the top 5% from the t distribution with $\delta = 20$?

   Answer: 1.725; use the 0.05 column and row 20 in Table A.2.

4. What is the top 5% from the χ^2 distribution with $\delta = 20$?

   Answer: 31.410; use the 0.05 column and row 20 in Table A.1.

5. What is the top 2.5% from the t distribution with $\delta = 3$?

Answer: 3.182; use the 0.025 column and row 3 in Table A.2.

6. What is the top 2.5% from the χ^2 distribution with $\delta = 3$?

   Answer: 9.348; use the 0.025 column and row 3 in Table A.1.

7. What is the top 1% from the t distribution with $\delta = 12$?

   Answer: 2.681; use the 0.01 column and row 12 in Table A.2.

8. What is the top 1% from the χ^2 distribution with $\delta = 12$?

   Answer: 26.217; use the 0.01 column and row 12 in Table A.1.

9. What is the top 5% from the t distribution with $\delta = 49$?

   Answer: 1.684; use the 0.05 column and row 40 in Table A.2.

   Note:: If the value of δ is not in the table, always go the next smaller value.

10. What is the bottom 5% from the t distribution with $\delta = 20$?

    Answer: -1.725; use the 0.05 column and row 20 in Table A.2.

    Note:: The t distribution is symmetric, so the negative is the bottom percentages. The χ^2 distribution is not symmetric, so using the negative does not work.

11. What is the bottom 2.5% from the t distribution with $\delta = 120$?

    Answer: -1.984; use the 0.025 column and row 100 in Table A.2.

9.4 Examples With R

In R, there are typically four functions for each probability distribution. The four functions start with a `d`, `q`, `p`, or `r`. The `d` stands for density, and the function returns the density value or the height of the curve. The `q` stands for quantile, and the function returns a quantile for a given probability. This is like looking up a probability in Table B.1 to find the associated z-score. The `p` stands for probability, and the function returns the probability for a given quantile. This is like finding a probability in Table B.1 for a given z-score. The `r` stands for random, and the function returns random values.

For the normal probability distribution the four functions are `dnorm`, `qnorm`, `pnorm`, and `rnorm`. For the t probability distribution the four functions are `dt`, `qt`, `pt`, and `rt`. For the χ^2 probability distribution the four functions are `dchisq`, `qchisq`, `pchisq`, and `rchisq`. Other probability distribution functions follow the same pattern. The use of these functions will be illustrated in the examples.

1. Find $P(N < 7)$ for a normal distribution with mean = 8 and sd = 3.

```
> pnorm(q=7, mean=8, sd=3)

[1] 0.3694413
```

2. Find $P(N > 89.3)$ for a normal distribution with mean = 87 and sd = 7.4.

```
> 1 - pnorm(q=89.3, mean=87, sd=7.4)

[1] 0.3779722

> ## the lower.tail=FALSE does a greater than probability

> pnorm(q=89.3, mean=87, sd=7.4, lower.tail=FALSE)

[1] 0.3779722
```

3. Find $P(-119.3 > N > -119.3)$ for a normal distribution with mean = -130.4 and sd = 17.7.

```
> pnorm(q=-129.3, mean=-130.4, sd=17.7) - pnorm(q=-119.3, mean=-130.4, sd=17.7)

[1] -0.2099322
```

4. Find x such that $P(N < x) = 0.674$ is true for a normal distribution with mean = 94 and sd = 19.4.

```
> qnorm(p=0.674, mean=94, sd=19.4)

[1] 102.7491
```

5. Find x such that $P(N > x) = 0.1674$ is true, for a normal distribution with mean = 153 and sd = 39.4.

```
> qnorm(p=1-0.1674, mean=153, sd=39.4)
[1] 191.0009
> qnorm(p=0.1674, mean=153, sd=39.4,lower.tail=FALSE)
[1] 191.0009
```

6. Generate 5 random values from a normal distribution with mean = 34 and sd = 9.4.

```
> set.seed(28389) #get the same random values after setting the seed
> rnorm(n=5, mean=34, sd=9.4)
[1] 38.95563 35.92742 29.15629 44.32952 18.00818
```

7. Return to the students' height scenario (page 78); only the calculations will be shown here. The answers will be slightly different because the tables in Appendix B are rounded.

```
> ## defined the mean and sd
> meanHeight <- 70
> sdHeight <- 2
```

(a) What is the probability that a male student is shorter than 68 inches?

```
> x <- 68
> pnorm(q=x,mean=meanHeight,sd=sdHeight)
[1] 0.1586553
```

(b) What is the probability that a male student is taller than 71 inches?

```
> x <- 71
> ## multiple ways to answer this
> 1-pnorm(q=x,mean=meanHeight,sd=sdHeight)
```

```
[1] 0.3085375
> pnorm(q=x,mean=meanHeight,sd=sdHeight,lower.tail=FALSE)
[1] 0.3085375
```

(c) What is the probability that a male student is between 67 and 72 inches?

```
> x1 <- 67
> x2 <- 72
> pnorm(q=x2,mean=meanHeight,sd=sdHeight)-pnorm(q=x1,mean=meanHeight,sd=sdHeight)
[1] 0.7745375
```

(d) What height corresponds to the 60th percentile?

```
> p <- 0.6
> qnorm(p=p,mean=meanHeight,sd=sdHeight)
[1] 70.50669
```

(e) What height corresponds to the top 10%?

```
> p <- 0.1
> ## multiple ways to solve this
> qnorm(p=1-p,mean=meanHeight,sd=sdHeight)
[1] 72.5631
> qnorm(p=p,mean=meanHeight,sd=sdHeight,lower.tail=FALSE)
[1] 72.5631
```

8. Return to class dismissal scenario (page 85); only the calculations will be show here. The answers will be slightly different because the tables in Appendix B are rounded.

```
> ## defined the mean and sd
> meanClass <- 50
> sdClass <- 4
```

(a) What is the top 20% for when class is dismissed?

```
> p <- 0.2

> ## there are several ways to do this
> qnorm(p=1-p,mean=meanClass,sd=sdClass)
[1] 53.36648
> qnorm(p=p,mean=meanClass,sd=sdClass,lower.tail=FALSE)
[1] 53.36648
```

(b) What is the probability that the class is let out between 8 minutes early and 8 minutes late?

```
> x1 <- 42
> x2 <- 58

> pnorm(q=x2,mean=meanClass,sd=sdClass) - pnorm(q=x1,mean=meanClass,sd=sdClass)
[1] 0.9544997
```

(c) What is the probability the class is dismissed 2 minutes late or sooner?

```
> x <- 52

> pnorm(q=x,mean=meanClass,sd=sdClass)
[1] 0.6914625
```

(d) What is the probability the class is dismissed 3 minutes early or later?

```
> x <- 47

> ## multiple ways to do this
> 1-pnorm(q=x,mean=meanClass,sd=sdClass)
[1] 0.7733726
> pnorm(q=x,mean=meanClass,sd=sdClass,lower.tail=FALSE)
[1] 0.7733726
```

(e) What is the 40th percentile for the time the class is dismissed?

```
> p <- 0.4
```

```
> qnorm(p=p,mean=meanClass,sd=sdClass)
[1] 48.98661
```

9. Find $P(t_7 < 0.34)$.

```
> pt(q=0.34, df=7)
[1] 0.6280846
```

10. Find $P(-0.234 < t_3 < 0.019)$.

```
> pt(q=0.019, df=3) - pt(q=-0.234, df=3)
[1] 0.09196061
```

11. Find $P(t_1 > -0.14)$.

```
> 1 - pt(q=-0.14, df=1)
[1] 0.5442756
> pt(q=-0.14, df=1, lower.tail=FALSE)
[1] 0.5442756
```

12. Generate 3 random values from a t_{103} probability distribution.

```
> set.seed(2827) #get the same random values after setting the seed
> rt(n=3, df=103)
[1]  0.3395757 -0.1083227 -0.3452918
```

13. Find $P(\chi^2_5 < 8.45)$.

```
> pchisq(q=8.45, df=5)
[1] 0.8668828
```

14. Find $P(3.75 < \chi^2_7 < 5.36)$.

```
> pchisq(q=5.36, df=7) - pchisq(q=3.75, df=7)
[1] 0.1919604
```

15. Find $P(\chi^2_2 > 4.7)$.

```
> 1 - pchisq(q=4.7, df=2)
[1] 0.09536916
> pchisq(q=4.7, df=2, lower.tail=FALSE)
[1] 0.09536916
```

16. Generate 4 random values from a χ^2_{10} probability distribution.

```
> set.seed(998) #get the same random values after setting the seed
> rchisq(n=4, df=10)
[1] 8.669138 5.713753 9.570490 7.573587
```

Chapter 10

Interval Estimation

In Chapter 7 the concept of the point estimate is developed. In particular the point estimate for μ, the population mean, is $\hat{\mu}$, the sample mean, and the point estimate for P, the population proportion, is $\hat{P}$, the sample proportion. The idea that the point estimate is not perfect and has some uncertainty was developed in Chapter 8, with the ideas of variance and standard deviation. The idea behind the interval estimate is to take into account the uncertainty with the point estimate. The interval estimate, or confidence interval (CI), as it is more commonly called, is based on three quantities: point estimate, uncertainty of the point estimate, and K multiplier. The K multiplier allows us to change how confident we are that our interval captured the true value of the parameter. The K multiplier comes from one of the probability distributions described in Chapter 9. Thus, the confidence interval combines and incorporates the ideas from Chapters 7, 8, and 9. In this text we will always calculate 95% confidence intervals. Confidence intervals will be calculated for the population mean, μ, and the population proportion, P.

10.1 Terms

Estimate: The statistical word for an educated guess.

Point estimate: A single number (point) for the educated guess.

Measure of error (MoErr): The amount of uncertainty associated with the point estimate.

K: The multiplier for the measure of error to change the level of confidence.

Interval estimate: A range (interval) of educated guesses (estimates). The interval estimate is also called a **confidence interval (CI)**. The interval estimate is the point estimate plus and minus K times the measure of error.

Outcome of interest: The outcome we use for the context of all parts of the interval estimation for proportions. The outcome of interest is determined by the question of interest.

Confidence and alpha level: The confidence level is a percentage describing how much confidence or surety there is in the interval estimate. The α level (or alpha level) is used to determine the K value. Confidence level $= (1 - \alpha)100\%$.

Recall from Chapter 7 the difference between μ and $\hat{\mu}$, and P and $\hat{P}$. The symbol with the "hat" is the point estimate for the symbol without the hat. Before we calculate a CI for a mean or a proportion, consider a generic example. For simplicity, in this text $\alpha = 0.05$. This means the confidence interval calculated will always have a 95% confidence level.

- θ is the population parameter.
- $\hat{\theta}$ is the point estimate for the population parameter.
- MoErr is the measure of error associated with $\hat{\theta}$.
- K is the multiplier needed to get a 95% CI.
- The CI is $\hat{\theta} \pm K * MoErr = (L = \hat{\theta} - K * MoErr, U = \hat{\theta} + K * MoErr)$.
- The CI is reported as (L, U), where L is called the lower bound and U is the upper bound of the CI.
- CI interpretation: **I'm 95% confident the interval (L, U) captures the true value of θ.**

We can use this generic example as a template to interpret any confidence interval. We just need to replace L and U with numbers and change θ to the context of the problem.

10.2 Interval for a Proportion

Recall that proportions come from a categorical variable. There are two methods to calculate a CI for a proportion: the **one sample proportion method** and the **media method**. The biggest difference is the measure of error (MoErr) for each method.

- P is the population proportion for the outcome of interest.

- $\hat{P}$ is the estimate for the population proportion, or the sample proportion for the outcome of interest.

One sample proportion method:

$$\hat{P} \pm 1.96\sqrt{\frac{\hat{P}(1-\hat{P})}{n}}$$

- MoErr $= \sqrt{\frac{\hat{P}(1-\hat{P})}{n}}$ is the measure of error for the one sample proportion CI method.
- $K = 1.96$ is the multiplier for the one sample proportion method to get a 95% CI. The value of 1.96 comes from the standard normal distribution (Chapter 9).

Media method:

$$\hat{P} \pm \frac{1}{\sqrt{n}}$$

- MoErr $= 1/\sqrt{n}$ is the measure of error for the media method. The media calls this quantity the **margin of error (MoE)**.
- $K = 1$ is the multiplier for the media method to get a 95% CI.

10.2.1 Examples

1. What proportion of college students own an Apple product? To answer this question, suppose 50 students were asked, "Do you own an Apple product: Yes or No?" $\alpha = 0.05$. Results: Yes (f = 30), No (f = 20).

   (a) Who is the population?

   Answer: All college students

   (b) What is the variable?

   Answer: Owning an Apple product or not

   (c) What are the outcomes?

   Answer: Yes or No

   (d) What is the outcome of interest?

   Answer: Yes. We are interested in students who own an Apple product.

   (e) What is the interpretation of P?

   Answer: The true proportion of college students who own an Apple product.

(f) What is the numerical value of $\hat{P}$?

Answer:

$$\hat{P} = \frac{30}{50} = 0.6$$

(g) Interpret the value of $\hat{P}$.

Answer: I predict that 60% of college students own an Apple product.

(h) Construct a 95% CI using the one sample proportion method.

Answer:

$$\hat{\theta} \pm K * MoErr \Rightarrow \hat{P} \pm 1.96\sqrt{\frac{\hat{P}(1-\hat{P})}{n}} \Rightarrow 0.6 \pm 1.96\sqrt{\frac{0.6(1-0.6)}{50}}$$
$$\Rightarrow 0.6 \pm 0.1358 \Rightarrow (0.6 - 0.1358,\ 0.6 + 0.1358) \Rightarrow (0.464,\ 0.735)$$

(i) Interpret the one sample proportion CI.

I'm 95% confident the interval (0.464, 0.735) captures the true proportion of college students who own an Apple product.

(j) Construct a 95% CI using the media method.

$$\hat{\theta} \pm K * MoErr \Rightarrow \hat{P} \pm 1 * \frac{1}{\sqrt{n}} \Rightarrow 0.6 \pm 0.1414 \Rightarrow (0.6 - 0.1414,\ 0.6 + 0.1414) \Rightarrow (0.459,\ 0.741)$$

(k) Interpret the media method CI.

I'm 95% confident the interval (0.459, 0.741) captures the true proportion of college students who own an Apple product.

What have we done? We have a point estimate and an interval estimate. I predict 60% of college students own an Apple product. I am 95% confident the true percentage of college students who own an Apple product is captured by the interval (46.4%, 73.5%) using the one sample proportion method or by the interval (45.9%, 74.1%) using the media method. So, if single number to guess (estimate) the proportion of college students who own an Apple product is desired, use $\hat{P}$. If a range of possible proportions is desired, use a CI.

2. Is Congress viewed favorably by American voters? To answer this question, suppose 1,111 American voters were randomly selected and asked, "Do you approve or disapprove of the actions of Congress?" $\alpha = 0.05$. Results: Approve (f = 354), Disapprove (f = 757).

(a) What is the question of interest?

Answer: Is Congress viewed favorably by American voters?

(b) Who is the population?

Answer: All American voters

(c) What is the variable?

Answer: Feelings about actions of Congress.

(d) What are the outcomes?

Answer: Approve or Disapprove

(e) What is the level of measurement?

Answer: Categorical

(f) What is the outcome of interest, and why?

Answer: Approve. We want to know if Congress is viewed favorably.

(g) What is the numerical value of $\hat{P}$?

Answer:

$$\hat{P} = \frac{354}{1111} = 0.32$$

(h) Interpret $\hat{P}$.

Answer: I predict that 32% of American voters approve of the actions of Congress.

(i) Calculate a 95% CI using the one sample proportion method.

Answer:

$$\hat{P} \pm 1.96\sqrt{\frac{\hat{P}(1-\hat{P})}{n}} \Rightarrow 0.32 \pm 1.96\sqrt{\frac{0.32(1-0.32)}{1111}} \Rightarrow (0.293, 0.347)$$

(j) Interpret the one sample proportion CI.

Answer: I'm 95% confident the interval (0.29, 0.35) captures the true proportion of American voters who approve of the actions of Congress.

(k) Construct a 95% CI using the media method.

Answer:

$$\hat{P} \pm \frac{1}{\sqrt{n}} \Rightarrow 0.32 \pm \frac{1}{\sqrt{1111}} \Rightarrow (0.29, 0.35)$$

(l) Interpret the media method CI.

Answer: I'm 95% confident the interval (0.29, 0.35) captures the true proportion of American voters who approve of the actions of Congress.

3. It is common for advertising to claim that 4 out of 5 experts endorse their product. For example, 4 out of 5 statisticians think confidence intervals are awesome. But does that imply the majority?

(a) What is the question of interest?

Answer: Do a majority of statisticians think confidence intervals are awesome?

(b) What is the variable?

Answer: Do you think confidence intervals are awesome?

(c) What are the outcomes?

Answer: Yes or No

(d) What is the level of measurement?

Answer: Categorical

(e) What is the outcome of interest?

Answer: Yes. A yes response is an endorsement of confidence intervals being awesome.

(f) What is the value of $\hat{P}$ and interpretation?

Answer:

$$\hat{P} = \frac{4}{5} = 0.8$$

I predict that 80% of statisticians think confidence intervals are awesome.

(g) Construct a 95% CI using the one sample proportion method.

Answer:

$$\hat{P} \pm 1.96\sqrt{\frac{\hat{P}(1-\hat{P})}{n}} \Rightarrow 0.8 \pm 1.96\sqrt{\frac{0.8(1-0.8)}{5}} \Rightarrow (0.45, 1.15)$$

Does it make sense that the upper bound is greater than 1? Is there a mistake? There is no mistake; sometimes the calculation for the CI for the proportion will produce negative values or values greater than 1.

Note: If the lower bound of the confidence interval for the **proportion** is negative, change the value to 0. If the upper bound of the confidence interval for the **proportion** is greater than 1, change the value to 1.

The CI becomes $(0.45, 1)$.

(h) Interpret the CI. I am 95% confident the interval $(0.45, 1)$ captures the true proportion of statisticians who think confidence intervals are awesome.

(i) Answer the question of interest.

Answer: No, the interval is not strictly above 0.5, so we are not confident that 50% or more of statisticians think confidence intervals are awesome.

4. The last example compared the CI to a target or goal value. The use of a CI in this way is very common. Consider the following fictitous example. The administration at the local university wants at least 60% of STEM majors to be female. The administration asked randomly selected STEM majors, "Are you female or male?"

(a) What is the outcome of interest?

Answer: Female. The outcome of interest for the administration.

(b) If $\hat{P} = 0.75$ and the 95% CI is (0.65, 0.85), should the administration be happy or unhappy? Explain.

Answer: Happy. The value 0.60 is below the interval. There is evidence that the proportion of female STEM majors is above 0.60.

(c) If $\hat{P} = 0.45$ and the 95% CI is (0.35, 0.55), should the administration be happy or unhappy? Explain.

Answer: Unhappy. The value 0.60 is above the interval. There is evidence that the proportion of female STEM majors is below 0.60.

(d) If $\hat{P} = 0.55$ and the 95% CI is (0.45, 0.65), should the administration be happy or unhappy? Explain.

Answer: Happy. The value 0.60 is inside the interval. There is *not* evidence that the proportion of female STEM majors is below 0.60.

The target value being inside the interval means it could be above or below; we are not sure. The target value being close to the lower or upper bound does not matter. It only matters if the target value is inside the interval or not.

10.2.2 Comparison of Proportion Interval Methods

Since there are two methods to calculate a confidence interval for a proportion, how do we choose between them? The methods will be compared through some questions:

1. In which interval method, one sample proportion or media, do we have more confidence?

   Answer: Both or neither. They have the same level of confidence, so we are equally confident in both of them.

2. Which interval method is better?

   Answer: That depends. In example 1 the one sample proportion CI is narrower than the media method. In example 2 the intervals are the same. What's the big difference between these two examples? The sample size. For a small sample size, the one sample proportion CI is better. For a large sample size, they are essentially the same. It is safe to say that one sample proportion is better.

3. Why do the media use $1/\sqrt{n}$ as the margin of error (MoE)?

   Answer: When a proportion (or percentage) is presented by the media, located at the bottom of the screen is almost always MoE $\pm$ 3%. This tells you approximately the value of the sample size. From example 2, $n = 1111$—this sample size gives the 3% MoE, which is essentially the same as the MoErr from the one sample proportion method. Since the media use large sample sizes and the media method is easier, that is what is used.

10.3 Interval for the Mean

Unlike the CI for a proportion, there is only one method for the CI for the population mean. Recall means are calculated from a ratio variable. The CI formula for the mean is

$$\hat{\mu} \pm t_{\delta,\alpha/2}\frac{S}{\sqrt{n}}$$

- μ is the population mean.
- $\hat{\mu}$ is the estimate for the population mean or the sample mean.
- MoErr $= \frac{S}{\sqrt{n}}$ is the measure of error for the CI on the mean. Recall that S is the standard deviation (Section 8.2.2).
- K $= t_{\delta,\alpha/2}$ is the "t" multiplier from Table A.2 in Appendix A. The "t" multiplier is based on δ (delta), the degrees of freedom, and the α level. Since $\alpha = 0.05$, $\alpha/2 = 0.025$. The degrees of freedom are $\delta = n - 1$. The "t" multiplier is looked up in Table A.2 in Appendix A. The $\alpha/2 = 0.025$ refers to the column to use and δ refers to which row to use. The "t" multiplier is named such because it comes from the t distribution (Section 9.3.1). The values in Table A.2 are the top 2.5% from the t distribution that are in the 0.025 column. Refer to Section 9.3.3 on meaning of the tables in Appendix A.

10.3.1 Examples

1. What is the mean number of apps college students have on their smartphone? To answer this question suppose 7 college students were asked, "How many apps do you have on your smartphone?" $\alpha = 0.05$. Results: $\hat{\mu} = 155.71$, $S = 22.65$.

   (a) Interpret μ.

   Answer: The true mean number of apps of all college students' smartphones.

   (b) Interpret $\hat{\mu}$.

   Answer: I predict the typical student has about 156 apps on their smart phone.

   (c) What is the numerical value for K (t multiplier)?

   Answer: First we need to know the degrees of freedom: $\delta = n - 1 = 7 - 1 = 6$.

   Using Table A.2 in Appendix A, go down to the row under δ to the number 6. Now go over to the column with the heading 0.025. This number is $t_{6,\alpha/2} = 2.447$.

   (d) Construct a 95% CI for the mean.

   Answer:

$$\hat{\theta} \pm K * MoErr \Rightarrow \hat{\mu} \pm t_{\delta,\alpha/2}\frac{S}{\sqrt{n}} \Rightarrow 155.71 \pm 2.447\frac{22.65}{\sqrt{7}} \Rightarrow (134.76, 176.66)$$

   (e) Interpret the CI.

Answer: We have the same problem again of how we can have a fractional app, so we round the lower bound of the interval down and the upper bound of the interval up. I'm 95% confident the interval (134, 177) captures the true mean number of apps on college students' smartphones.

2. The CEO of a major fictitious company wants to know if the employees are spending more than 3 hours per day on Facebook while at work. To answer this question the computer activity of 40 employees was monitored and the number of hours in a day they were on Facebook was recorded. Results: mean = 4.5, standard deviation = 3.2.

   (a) What is the question of interest?

   Answer: Do employees spend more than 3 hours a day on Facebook while at work?

   (b) Who is the population?

   Answer: All the employees of the company.

   (c) What is the variable?

   Answer: Hours spent on Facebook.

   (d) What is the level of measurement?

   Answer: Ratio

   (e) What is the t multiplier for the CI?

   Answer: $t_{35,0.025} = 2.030$

   $\delta = 40 - 1 = 39$. That value is not in Table A.2. Therefore, use the next lower value of $\delta = 35$.

   Note: If a calculated value of δ does not appear in Table A.2, use the next smaller value.

   (f) Construct the 95% CI for the mean.

   Answer:

$$\hat{\mu} \pm 2.030 \frac{S}{\sqrt{n}} \Rightarrow 4.5 \pm 2.030 \frac{3.2}{\sqrt{40}} \Rightarrow (3.473,\ 5.527)$$

   (g) Interpret the CI.

   Answer: I'm 95% confident the interval (3.47, 5.53) captures the true mean number of hours spent on Facebook by employees at work.

(h) Answer the question of interest.

Answer: There is evidence that employees spend more than 3 hours per day on Facebook while at work. The CI is completely above the value of 3 hours.

3. Fred (fictitious) considers himself unlucky. He recorded his winning amounts from his last 90 games of roulette. Roulette is a gambling game in which participants can either win or lose money. Results: mean = -1.5; standard deviation = 10.2

(a) Interpret $\hat{\mu}$.

Answer: I predict that Fred will lose \$1.5 the next time he plays roulette.

(b) Calculate a 95% CI for the mean.

Answer: $\delta = 90 - 1 = 89$; that value is not in the table, so use 80.

$$\hat{\mu} \pm t_{\delta,0.025}\frac{S}{\sqrt{n}} \Rightarrow -1.5 \pm 1.99\frac{10.2}{\sqrt{90}} \Rightarrow (-3.64, 0.64)$$

(c) Interpret the CI.

Answer: I am 95% confident the interval (-3.64, 0.64) captures the true mean amount of Fred's winnings from playing roulette.

Note: The negative lower bound is *not* changed to zero. A negative value in this context is interpreted as losing money.

(d) Answer the question of interest.

Answer: Fred should not consider himself unlucky. The upper bound of the CI is positive. We do not have evidence that Fred always loses at roulette.

10.4 Examples With R

When calculating a CI, it is a great time to also practice function writing (Section 4.7). Here is a function that calculates the CI for a proportion using the one sample proportion method:

```
> oneSamplePropCI <- function(count, n){
    ## count is the frequency for the outcome of interest
    ## n is the sample size
```

```
    k <- qnorm(0.975) # better than the rounded value of 1.96
    p <- count/n # p hat, the estimated proportion
    L <- p - k*sqrt(p*(1-p)/n) # lower bound
    U <- p + k*sqrt(p*(1-p)/n) # upper bound
    out <- c(Phat=p, L=L, U=U) # named vector of p hat and the bounds
    return(out) # return the interval
}
```

Here is a function that calculates the CI for a proportion using the media method:

```
> mediaCI <- function(count, n){
    ## count is the frequency for the outcome of interest
    ## n is the sample size
    k <- 1 #not needed but good to remember
    p <- count/n # p hat, the estimated proportion
    L <- p - k*1/sqrt(n) # lower bound
    U <- p + k*1/sqrt(n) # upper bound
    out <- c(Phat=p, L=L, U=U) # named vector of p hat and the bounds
    return(out) # return the interval
}
```

Here is a function that calculates the CI for a mean using the sample mean and standard deviation:

```
> meanCI <- function(mean, sd, n){
    ## mean = sample mean, sd = standard deviation, n = sample size
    t <- qt(p=0.975, df = n-1) # t multiplier
    L <- mean - t*sd/sqrt(n) # lower bound
    U <- mean + t*sd/sqrt(n) # upper bound
    return(c(L,U)) # return the vector in one step
}
```

The results may be a little different because less rounding is occuring in R. Also, there is no need to use Table A.2, meaning the exact value of δ can be used.

1. Return to the Apple product scenario (page 103):

```
> oneSamplePropCI(30, 50)
     Phat         L         U
0.6000000 0.4642097 0.7357903
> mediaCI(30, 50)
     Phat         L         U
0.6000000 0.4585786 0.7414214
```

2. Return to the Congress approval scenario (page 104):

```
> oneSamplePropCI(354, 1111)
     Phat         L         U
0.3186319 0.2912334 0.3460303
> mediaCI(354, 1111)
     Phat         L         U
0.3186319 0.2886304 0.3486334
```

3. Return to the smartphone app scenario (page 109); assume the enumeration is

   166 119 147 185 153 142 178

   There is another function to calculate the CI from the enumeration directly. The function produces other output, but we care only about the CI right now:

```
> apps <- c(166, 119, 147, 185, 153, 142, 178)
> t.test(x=apps, conf.level=0.95)$conf.int
[1] 134.7621 176.6664
attr(,"conf.level")
[1] 0.95
```

4. Return to the employees on Facebook scenario (page 110):

```
> meanCI(4.5, 3.2, 40)
[1] 3.47659 5.52341
```

Chapter 11

Hypothesis Testing

Hypothesis testing is the foundation of science. In science a new theory is proposed, and that theory is assessed (tested) against what is observed (samples, variables, enumeration, etc.). The majority of the text before this chapter laid the foundation for explaining hypothesis testing. The remaining chapters are specific hypothesis testing situations. This chapter covers the language of hypothesis testing and provides an overview of it.

11.1 Terms

Hypothesis: A speculative statement about the population.

Null hypothesis (H0): A statement about the population we are willing to assume is true.

Alternative hypothesis (HA): A statement about the population we would like to demonstrate.

Test statistic: The amount of evidence in the sample that supports the alternative hypothesis.

Critical value: The minimum amount of evidence needed to abandon the null hypothesis and conclude the alternative hypothesis. At this point it is unreasonable to keep holding to our assumption that the null hypothesis is true.

Significant: The decision to reject the null hypothesis.

Type I error: To conclude the alternative hypothesis when in fact the null hypothesis is true.

Type II error: To conclude the null hypothesis when in fact the alternative hypothesis is true.

Alpha (α): The probability of a type I error. The probability of concluding the alternative hypothesis when in fact the null hypothesis is true.

Beta (β): The probability of a type II error. The probability of concluding the null hypothesis when in fact the alternative hypothesis is true.

Level of confidence: The probability of concluding the null hypothesis when in fact the null hypothesis is true.

Power: The probability of concluding the alternative hypothesis when in fact the alternative hypothesis is true.

Null distribution: A probability distribution (Chapter 9) used for hypothesis testing.

P-value: The probability of being more extreme than the test statistic using the null distribution.

The terms "alpha" and "level of confidence" were introduced in Chapter 10. These terms are basically the same, just defined in the context of confidence intervals or hypothesis testing.

11.2 Hypothesis Testing Overview

11.2.1 Hypotheses

Hypothesis testing begins with a question, denoted as the question of interest (Chapter 1). The hypotheses are statements about the population that are possible answers to the question of interest. The null hypothesis is a statement that is assumed to be true. If no observations were made or there was not enough information collected, the null hypothesis is the best guess to the question of interest. The alternative hypothesis is a statement about the population, the answer to the question of interest that we are trying to demonstrate. The new theory in science is represented in the alternative hypothesis. The hypotheses are a translation from the language of a scientific discipline to the language of statistics.

11.2.2 Test Statistic

After the hypotheses are determined, information is collected to test the hypotheses. The collected information is in terms of variables (Chapter 2), samples (Chapter 3), point estimates (Chapter 7), and variability (Chapter 8). The collected information is summarized into the test statistic. The test statistic is the amount of evidence in the sample that supports the alternative hypothesis.

11.2.3 Statistical Decision

A statistical decision can be made based on the collected information (test statistic value). The test statistic is compared to the critical value or the p-value is compared to alpha to make the statistical decision. In statistics there are two possible decisions:

- Fail to reject the null hypothesis.
- Reject the null hypothesis.

The decisions sound a little strange. This goes with how hypothesis testing is designed. If we fail to reject the null hypothesis, this means we didn't have enough evidence in our sample for the alternative hypothesis. This does not prove the null hypothesis is true; there is not enough evidence to disprove the null hypothesis. If we reject the null hypothesis, this means we did have enough evidence in our sample for the alternative hypothesis. There is enough evidence to disprove the null hypothesis

11.2.4 Conclusion/Interpretation

The conclusion or interpretation of the statistical decision is the translation back from the language of statistics to the scientific discipline. If the statistical decision is fail to reject the null hypothesis the conclusion, is to remain with the null hypothesis. The null hypothesis is what is initially assumed to be true. Not enough evidence was in the collected information to reject null hypothesis, thus we still hold to it. If the statistical decision is reject the null hypothesis, then we conclude the alternative hypothesis. Enough evidence was in the collected information to abandoned the original assumption, and therefore, the alternative hypothesis is concluded.

11.2.5 Court Case Analogy

In the U.S. court system, when someone is put on trial they are assumed innocent until proven guilty. The **null hypothesis** is "the defendant is innocent." This is the statement we are assuming to be true from the beginning. The **alternative hypothesis** is "the defendant is guilty." This is the statement trying to be demonstrated. The **test statistics** is the evidence provided by the prosecutors that supports the claim the defendant is guilty. The **critical value** is the point of being beyond a reasonable doubt in the court case. After the evidence has been given, the jury makes one of two decisions: not guilty or guilty. Notice there is no decision of innocent. The jury's decision is like the statistical decision. The "not guilty" is like failing to reject the null hypothesis. There is not sufficient evidence to show guilt (or the alternative hypothesis). The "guilty" decision is like rejecting the null hypothesis. After the jury's decision, how the person is treated is the conclusion or interpretation of the decision. A "not guilty" verdict does not prove the person is innocent; however, society treats them as innocent. The person does not lose any rights and does not face any consequences. A "guilty" verdict mean the person is treated as having committed the crime. The person is sentenced to consequences (prison, fine, etc.). Society treats this person differently and not as innocent any more.

11.3 Conclusion Table

The legal system is not perfect, and neither is statistics, and errors are made. In statistics these errors are given names and have probabilities associated with them. Table 11.1 presents the conclusion table. The columns represent our conclusions and the rows are the actual truth.

		Conclusion	
		H0 is true HA is false	HA is true H0 is false
Actual Truth	H0 is true HA is false	Correct Decision Level of Confidence $1-\alpha$	Incorrect Decision Type I Error α Level of Significance
	HA is true H0 is false	Incorrect Decision Type II Error β	Correct Decision Power $1-\beta$

Table 11.1: Hypothesis testing conclusion reference table.

- The first row and first column conclude the H0 is true, when in fact this is true. This is a correct conclusion. The level of confidence is the probability of concluding the H0, when in fact the H0 is true. The probability is denoted as $1-\alpha$ (pronounced 1 minus alpha).
- The first row and second column conclude the HA is true, when in fact the opposite is true. This is an incorrect conclusion, which is called a type I error. A type I error is when a researcher concludes the HA is true when in fact the H0 is true. The level of significance is the probability of a type I error. The level of significance is the probability of concluding the HA is true, when in fact H0 is true. The level of significance is given the symbol α (alpha).
- The second row and first column conclude the H0 is true, when in fact the HA is true. This is an incorrect conclusion called a type II error. A type II error is when a researcher concludes the H0 is

true, when in fact the HA is true. The probability of a type II error is given the symbol β (beta). Therefore, β is the probability of concluding the H0 is true, when in fact the HA is true.

- The second row and second column conclude the HA is true, when in fact the HA is true. This is a correct conclusion. The probability of this correct conclusion is called power. Power is the probability of concluding the HA is true, when in fact the HA is true. Power is given the symbol $1 - \beta$ (1 minus beta).

If we know the truth, why commit an error? Or, how do we know we are committing an error? The answer is that, in practice, we do not know the truth and do not know if an error has been committed. It is important to understand that errors can happen.

As previously mentioned in this text, $\alpha = 0.05$. This means the probability of a type I error is set to 0.05. Or, there is a 5% chance of a type I error. This also means the level of confidence is $0.95 = 1 - \alpha$. There is a 95% chance of concluding the H0 when the H0 is true.

The value of the probability for power and β cannot be determined from α. How to calculate power and β is beyond the scope of this text. However, it should be understood that power being close to 1 and β being close to 0 is desirable. In subsequent statistic classes/texts, the conclusion table is discussed in more detail.

11.4 Null Distribution

As previously described (Section 11.2), hypothesis testing is the process of making a decision with two outcomes: Reject the null hypothesis or fail to reject the null hypothesis. This decision is based on samples and statistics, which are surrogates for populations and parameters (Chapter 3). Because of this, there is uncertainty involved in the decision. The uncertainty is quantified using a probability distribution. The null distribution is the distribution used for probability statements. The exact distribution used is different for each hypothesis testing situation. The potential distributions used as null distributions are described in Chapter 9.

The null distribution is typically used in one of two ways: determine the critical value or determine the p-value. The p-value is a probability. If the null distribution is the normal distribution, then the p-value can be looked up in a table (Section 9.2). If the null distribution is another probability distribution, the p-value can only be calculated with a computer. For this reason critical values are used, because they can be looked up in a table and do not require the use of a computer. In the scientific literature and in most statistical analyses, the use of the p-value is much more common. This text will use both critical values, for the reader not using statistical software, and p-values. A reader not using statistical software should still understand the concept of a p-value.

11.4.1 Critical Value

The critical value (CV) is the minimum amount of evidence needed to reject the null hypothesis. The CV comes from the null distribution using a particular probability statement. The probability statement is something like finding the top 5% or the bottom 5%. The exact nature of the probability statement depends on the hypothesis testing situation and the alternative hypothesis.

If the CV is to be used, then typically no p-value is calculated. By setting $\alpha = 0.05$, which is always the case for this text, this also sets the type I error. It also sets the CV (the minimum amount of evidence) that is required to reject the null hypothesis. The CV is compared to the test statistic (TS) to determine the statistical decision:

- **Reject the null hypothesis** if the TS is more extreme than the CV.
- **Fail to reject the null hypothesis** if the TS is not more extreme than the CV.

The definition of "more extreme" is different for each specific hypothesis testing situation.

11.4.2 P-Value

The p-value is a probability calculated from the null distribution. The p-value is the probability of being equal to or more extreme than the TS value. "More extreme" can be a greater than or less than probability depending on the alternative hypothesis. The exact nature of the probability statement depends on the hypothesis testing situation and the alternative hypothesis.

If the p-value is calculated, the critical value is typically not used. The p-value is compared to α to make the statistical decision:

- **Reject the null hypothesis** if the p-value $< \alpha$.
- **Fail to reject the null hypothesis** if the p-value $> \alpha$.

This process has an added advantage: The p-value can be reported, and the reader can apply their own α value.

11.5 Steps of Hypothesis Testing

There are several steps to hypothesis testing. The exact number is not important, and other sources have more or fewer steps than those listed, because steps are combined or separated. The order of the step is what is important. The steps must be done, in order, and typically previous steps dicate what to do at each step. There are many different hypothesis testing situations that use all these steps. However, the specifics of each step are a little different for each situation:

1. Identify the **question of interest**. This is not exactly statistics, but the point of hypothesis testing is to answer this question. Any answer that does not match this question is useless.

2. Identify the **appropriate test situation** and specify α. For this text $\alpha = 0.05$, always. To determine the test situation three questions need to be answered:

   (a) How many samples are there?

   (b) How many variables are there?

   (c) What is the level of measurement?

3. Determine the **null hypothesis (H0)** and **alternative hypothesis (HA)**. The question of interest is where to look to form these statements about the population, in particular, the HA.

4. **Select the critical value (CV)** and determine the equation of the test statistic. The HA helps in determining the critical value, but it is different for every test situation. The test statistic is some function of the summary statistics, which is different for every test situation. If statistical software is going to be used, then the CV is omitted. Instead of determining the equation of the test statistics, determine how to calculate the test statistic using the software.

5. Collect the sample, measure the variables, and **calculate the test statistic (TS)**. Collecting the sample from a population was discussed in Chapter 3. Calculating the test statistic is done using the summary statistics, for example, the frequency, proportion, or mean. The test statistic is different for each test situation. If statistical software is being used, then also calculate the p-value.

6. Make the **statistical decision**. The statistical decision can be based on the TS and CV (Section 11.4.1) or the the p-value and α (Section 11.4.2). In practice the most common approach is the p-value, but statistical software is usually required. It is very uncommon to do both. However, as long as α does not change between approaches, the decision will be the same.

7. **Conclusion, interpret the statistical decision, or answer the question of interest.** This is stating, in the context of the question of interest, what it means to reject the H0 or fail to reject the H0. The conclusion is done in slightly different ways for each hypothesis testing situation.

Figure 11.1 presents the steps of hypothesis testing in a flowchart format. The flowchart emphasizes the order and connection of the steps. Every problem starts with a scenario, from which we gain the information required. The two circles (scenario and enumeration) are what is provided in the word problem scenario. The rectangles are all the pieces to be determined from completing the hypothesis testing. To determine one of the rectangles, a rectangle (or circle) leading into that rectangle needs to first be determined. For example, before the hypotheses can be determined, the test situation and the questions of interest need to be determined. The dotted horizontal line at the enumeration circle indicates that no collected information (enumeration) is required to determine the rectangles above that line.

11.6 Test Situations Overview

The rest of this text will be devoted to different test situations. To determine which test situation to use, three questions *must* be answered:

1. How many samples are in the scenario?
2. How many variables are in the scenario?
3. What is the level of measurement for each variable mentioned?

Table 11.2 presents the answers for each test situation. The answers are used as a guide to determine the correct hypothesis testing situation. It appears that one sample proportion and multinomial have the same answers to the three questions. This is true; the difference is the number of outcomes. This is the only case, these two test situations, that the number of outcomes matters in making a determination.

Test Situation	Number of Samples	Number of Variables	Level of Measurement
One Sample Proportion	1	1	Categorical (only two outcomes)
Multinomial	1	1	Categorical (three or more outcomes)
Pre-Post (Change)	1	1	Ratio
Homogeneity	2	1	Categorical
Two Independent Samples t	2	1	Ratio
Correlation	1	2	Ratio
Independence	1	2	Categorical

Table 11.2: How to determine the appropriate hypothesis testing situation.

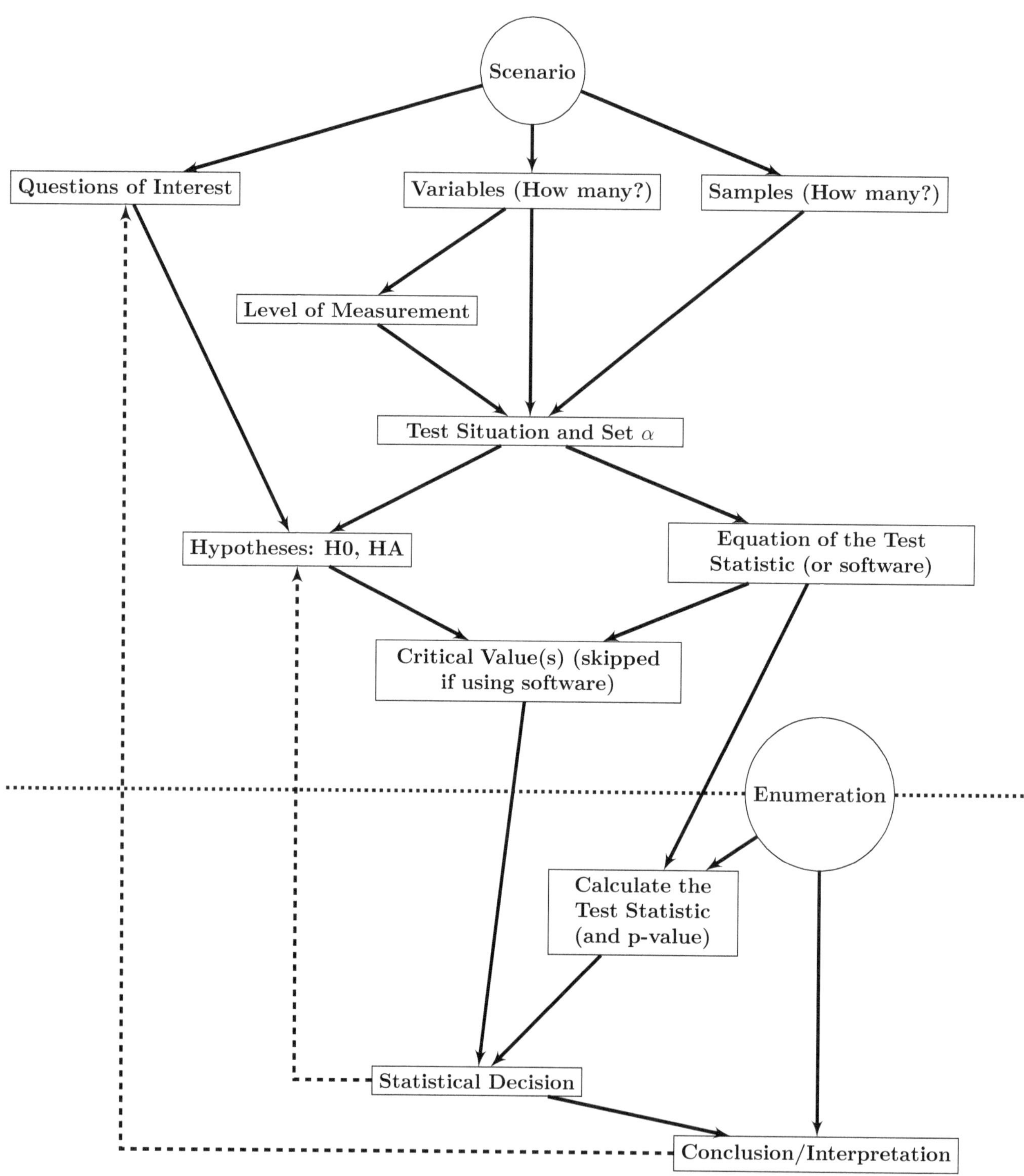

Figure 11.1: Hypothesis testing flowchart. Circles represent information provided in the scenario. Rectangles represent what should be determined using the steps of hypothesis testing. No enumeration or summary statistics are required to determine the rectangles above the horizontal dotted line at the enumeration circle.

11.7 Reading a Scenario

For the rest of this text, scenarios (word problems) will be presented. The ability to read these scenarios and extract the relevant pieces of information is the hardest part. Every scenario has four main pieces and one extra piece of information:

- The questions of interest
- The sample or samples
- The variable or variables
- The results, in terms of the enumeration or summary statistics
- Filler information to transition the other pieces within a paragraph

The question of interest can sometimes be a claim and not a question. It is usually either stated as a claim by some party or as a question asked of you, the student, reading the scenario. The interpretation/conclusion should answer this question or address the claim made. Typically, the sample(s) and variable(s) are in the same sentence. The variable is what is asked of the sample, not of you the student reading the scenario. The results are usually given at the end of the scenario. The filler is just extra information to help the scenario paragraph make sense.

11.8 Examples

1. The local city council (fictitious) is conducting a study to determine residents' opinions about parking downtown. They want to know if residents are in favor of installing parking meters downtown. The city council is willing to assume that residents are in favor of installing parking meters downtown unless the opposite can be demonstrated.

   (a) What is the H0?

   Answer: The local residents are in favor of installing parking meters downtown.

   (b) What is the HA?

   Answer: The local residents are not in favor of installing parking meters downtown.

   (c) What is the type I error?

   Definition: Concluding the HA when the H0 is true.

Answer: Concluding the local residents are not in favor of support installing parking meters downtown, when in fact the local residents are in favor of installing parking meters downtown.

(d) What is α?

Definition: $\alpha = 0.05$ *and* the probability of concluding the HA when the H0 is true.

Answer: The probability of concluding the local residents are not in favor of installing parking meters downtown, when in fact the local residents are in favor of installing parking meters downtown.

(e) What is the type II error?

Definition: Concluding the H0 when the HA is true.

Answer: Concluding the local residents are in favor of installing parking meters downtown, when in fact the local residents are not in favor of installing parking meters downtown.

(f) What is β?

Definition: The probability of concluding the H0 when the HA is true.

Answer: The probability of concluding the local residents are in favor of installing parking meters downtown, when in fact the local residents are not in favor of installing parking meters downtown.

(g) What is power?

Definition: The probability of concluding the HA when the HA is true.

Answer: The probability of concluding the local residents are not in favor of installing parking meters downtown, when in fact that is true.

(h) What is level of confidence?

Definition: The probability of concluding the H0 when the H0 is true.

Answer: The probability of concluding the local residents are in favor of installing parking meters downtown, when in fact that is true.

(i) What is the CV?

Definition: The minimum amount of evidence required to reject the H0.

Answer: The minimum amount of evidence required to reject that that local residents are in favor of installing parking meters downtown.

(j) What is the TS?

Definition: The amount of evidence collected supporting the HA.

Answer: The amount of evidence collected supporting that the local residents are not in favor of installing parking meters downtown.

(k) What is the p-value?

Definition: The probability of being more extreme than the TS.

Answer: The probability of more extreme evidence collected that supports that the local residents are not in favor of installing parking meters downtown, assuming the H0 is true.

2. The U.S. House of Representatives (fictitious) is debating on whether to increase the federal income tax. They believe that the American people are in favor of an increase in income tax, because it will help reduce the national deficit. The House of Representatives decided to spend $4.5 trillion on a study to determine if the American people are actually not in favor of an increase in income tax.

Try to answer the questions first before looking at the answers:

(a) What is the H0?

(b) What is the HA?

(c) What is the type I error?

(d) What is α?

(e) What is the type II error?

(f) What is β?

(g) What is power?

(h) What is level of confidence?

(i) What is the CV?

(j) What is the TS?

(k) What is the p-value?

(a) What is the H0?

Answer: The American people are in favor of an increase in federal income tax.

(b) What is the HA?

Answer: The American people are not in favor of an increase in federal income tax.

(c) What is the type I error?

Answer: Concluding the American people are not in favor of an increase in federal income tax, when in fact the American people are in favor of an increase in federal income tax.

(d) What is α?

Answer: The probability of concluding the American people are not in favor of an increase in federal income tax, when in fact the American people are in favor of an increase in federal income tax.

(e) What is the type II error?

Answer: Concluding the American people are in favor of an increase in federal income tax, when in fact the American people are not in favor of an increase in federal income tax.

(f) What is β?

Answer: The probability of concluding the American people are in favor of an increase in federal income tax, when in fact the American people are not in favor of an increase in federal income tax.

(g) What is power?

Answer: The probability of concluding the American people are not in favor of an increase in federal income tax, when that is true.

(h) What is level of confidence?

Answer: The probability of concluding the American people are in favor of an increase in federal income tax, when that is true.

(i) What is the CV?

Answer: The minimum amount of evidence required to reject that the American people are in favor of an increase in federal income tax.

(j) What is the TS?

Answer: The amount of evidence collected supporting that the American people are not in favor of an increase in federal income tax.

(k) What is the p-value?

Answer: The probability of the evidence collected or more extreme that supports the American people are not in favor of an increase in federal income tax.

3. Which university has smarter students, UW or UW? Suppose 50 randomly selected students from the University of Wyoming and another 60 randomly selected students from the University of Washington were studied. From each student it was recorded if they are on the honor roll: Yes or No.

   (a) What is the question of interest?

   Answer: Which university has smarter students, UW or UW?

   (b) How many variables, and what are they?

   Answer: One variable: whether the student is on the honor roll.

   (c) What is the level of measurement?

   Answer: Categorical. The outcomes Yes and No cannot be ordered.

   (d) How many samples, and what are they?

   Answer: Two samples: 50 Wyoming students and 60 Washington students.

   (e) What is the test situation? Provide justification.

   Answer: Homogeneity

   - Two samples: 50 Wyoming students, 60 Washington students
   - One variable: Whether the student is on the honor roll
   - Categorical level of measurement

4. Which university has smarter students, UW or UW? Suppose 50 randomly selected students from the University of Wyoming and another 60 randomly selected students from the University of Washington had their GPA recorded.

   (a) What is the question of interest?

   Answer: Which university has smarter students, UW or UW?

   (b) How many variables, and what are they?

   Answer: One variable: GPA

   (c) What is the level of measurement?

   Answer: Ratio

   (d) How many samples, and what are they?

   Answer: Two samples: 50 Wyoming students and 60 Washington students

(e) What is the test situation? Provide justification.

Answer: Two independent samples t

- Two samples: 50 Wyoming students and 60 Washington students
- One variable: GPA
- Ratio level of measurement

5. It has been hypothesized that the consumptions of coffee in the evening before going to bed can be adversely related to your ability to sleep. To examine this claim, suppose 20 graduate students volunteered to participate in a study and were asked two questions:

- How much coffee (measured in ounces) did you drink last night before going to bed?
- How many hours of sleep did you get last night?

(a) What is the question of interest?

Answer: The consumption of coffee in the evening before going to bed can be adversely related to your ability to sleep.

(b) What is the appropriate test situation? Provide justification.

Answer: Correlation

- One sample: 20 graduate students
- Two variables: Amount of coffee consumed and hours of sleep
- Both are ratio level of measurement

6. It has been hypothesized that the consumptions of coffee in the evening before going to bed can be adversely related to your ability to sleep. To examine this claim, suppose 20 graduate students volunteered to participate in a study and were asked two questions:

- Did you drink coffee last night before going to bed: Yes or No?
- Rate the quality of your sleep last night: Poor, Fair, or Great?

(a) What is the question of interest?

Answer: The consumption of coffee in the evening before going to bed can be adversely related to your ability to sleep.

(b) What is the appropriate test situation? Provide justification.

Answer: Independence

- One sample: 20 graduate students
- Two variables: Was coffee consumed and quality of sleep
- Both are categorical levels of measurement (quality of sleep is ordinal, treated as categorical)

7. Are basketball shoes equally or unequally preferred by NBA players? To test this claim, suppose 75 NBA players were asked, "Which brand of basketball shoe do you prefer: Adidas (A), Nike (N), Reebok (R)?" Results: A (f = 32), N (f = 21), R (f = 22).

   (a) What is the question of interest?

   Answer: Are basketball shoes equally or unequally preferred by NBA players?

   (b) What is the appropriate test situation? Provide justification.

   Answer: Multinomial

   - One sample: 75 NBA players
   - One variable: Which brand of basketball shoe do you prefer?
   - Categorical level of measurement
   - Three or more outcomes: Adidas, Nike, Reebok

8. Are Nike shoes preferred by NBA players? To test this claim suppose 75 NBA players were asked, "Do you prefer Nike: Yes or No?" Results: Yes (f = 44), No (f = 31).

   (a) What is the question of interest?

   Answer: Are Nike shoes preferred by NBA players?

   (b) What is the appropriate test situation? Provide justification.

   Answer: One sample proportion

   - One sample: 75 NBA players
   - One variable: Do you prefer Nike?
   - Categorical level of measurement
   - Only two outcomes: Yes or No

9. Does taking an introductory statistics class help university students understand the concept of correlation? Suppose an instructor gave Intro Statistics students a quiz about correlation at the beginning of the semester. During finals week, the students took the exact same quiz. Each time the percent grade was recorded.

   (a) What is the question of interest?

   Answer: Does taking an introductory statistics class help university students understand the concept of correlation?

   (b) What is the appropriate test situation? Provide justification.

   Answer: Pre-post (change)

   - One sample: Intro Statistics students
   - One variable: Correlation quiz, measured twice
   - Ratio level of measurement
   - Intervention: Taking an Intro Statistics course

Chapter 12

One Sample Proportion

The one sample proportion test situation is when there is one sample, one categorical variable, and only two outcomes. The purpose of this test situation is to compare the two outcomes to each other or to compare the outcome of interest to a target or comparison group. Since there are only two outcomes, knowing the proportion of the outcome of interest gives all of the required information, along with the sample size.

12.1 Terms

One sample proportion test: One sample with one categorical variable, with only two outcomes.

Outcome of interest: The outcome used for the context of all parts of the test situation.

P: The proportion of the outcome of interest for the whole population.

$\mathbf{P_0}$: The assumed value for proportion of the outcome of interest.

$\hat{\mathbf{P}}$: The estimate proportion of the outcome of interest; or the relative frequency for the outcome of interest.

n: The sample size.

Z: Standard normal distribution (Section 9.2.2).

12.2 Template

Once it has been determined that the test situation is one sample, here is how to conduct the test:

Outcome of interest:

1. Which outcome is in the context of the question of interest?
2. If neither outcome is specifically indicated, then choose the outcome that is listed first in the question of interest.

Value of $\mathbf{P_0}$:

1. Is there a reference group we are comparing to the sample?
   If yes, then the indicated value of the reference group is the P_0 value.
2. Does the question of interest ask about achieving a specific goal or target?
   If yes, then the specific value of the goal is the P_0 value.
3. If no to questions 1 and 2, then $P_0 = 0.5$.

H0: The population proportion equals P_0. H0: $P = P_0$.

Alternative hypothesis: One of the following based on the question of interest:

- HA: The population proportion is greater than P_0. HA: $P > P_0$.
- HA: The population proportion is less than P_0. HA: $P < P_0$.
- HA: The population proportion is not equal to P_0. HA: $P \neq P_0$.

Equation of the test statistic:

$$Z = \frac{\hat{P} - P_0}{\sqrt{\frac{P_0(1-P_0)}{n}}}$$

Null distribution: Z or the standard normal distribution (Section 9.2.2)

Critical value(s): The HA determines which CV to use from the Z distribution:

- CV = 1.645 if the HA is greater than (>)
- CV = -1.645 if the HA is less than (<)
- CV = -1.960 and 1.960 if the HA is not equal to ($\neq$)

Test statistic: Fill the values of $\hat{P}$, P_0, and n into

$$TS = \frac{\hat{P} - P_0}{\sqrt{\frac{P_0(1-P_0)}{n}}}$$

p-value: The probability statement for the p-value depends on the HA:

- If HA: $P > P_0$, p-value $= P(Z > \text{TS})$.
- If HA: $P < P_0$, p-value $= P(Z < \text{TS})$.
- If HA: $P \neq P_0$, p-value $= 2P(Z > |\text{TS}|)$.

Statistical decision:

- Fail to reject the H0, if p-value $> \alpha$ or
  - if TS < 1.645 and HA: $P > P_0$
  - if TS > -1.654 and HA: $P < P_0$
  - if TS > -1.96 and TS < 1.96 and HA: $P \neq P_0$
- Reject the H0, if p-value $< \alpha$ or
  - if TS > 1.645 and HA: $P > P_0$
  - if TS < -1.654 and HA: $P < P_0$
  - if TS is < -1.96 or TS > 1.96 and HA: $P \neq P_0$

Conclusion/interpretation:

- If the statistical decision is to fail to reject the H0, then conclude the H0 in English.
- If the statistical decision is to reject the H0, then conclude the HA in English and interpret the estimated proportion ($\hat{P}$).

12.3 Examples

As mentioned previously, when conducting a hypothesis test in practice, either the CV or the p-value is used for the statistical decision, not both. The examples will demonstrate both for completeness.

1. Do college students want more campus parking? To answer this question, suppose 50 randomly selected students were asked, "Should the trustees approve the creation of a new campus parking lot: Yes or No?" Results: Yes (f = 38), No (f = 12). $\alpha = 0.05$.

   Question of interest: Do college students want more campus parking?

   Number of samples: There is one sample: the 50 students.

Number of variables: There is one variable: "Should the trustees create a new parking lot?"

Level of measurement: Categorical.

Number of outcomes: Two: Yes and No.

Test situation: One sample proportion

- One sample: 50 students
- One variable: "Should the trustees create a new parking lot?"
- Categorical with only two outcomes

Outcome of interest: Yes. "Yes" being in favor of the creation of a new parking lot is in the context of the question of interest.

Interpret P: The proportion of all students who are in favor of the creation of a new campus parking lot.

$\mathbf{P_0}$: There is no reference group or specified goal; $P_0 = 0.5$.

Interpret $\mathbf{P_0}$: 50% of students are in favor of creating a new campus parking lot.

H0: $P = 0.5$, in English: The proportion of students who are in favor of the new parking lot is **equal** to the proportion of students who are not in favor of the new parking lot.

HA: $P > 0.5$, in English: The proportion of students who are in favor of the new parking lot is **greater than** the proportion of students who are not in favor of the new parking lot.

The question of interest is asking about "more campus parking." This is asking if more students will respond Yes to the variable than respond No. This is why the HA is "greater than."

Equation of test statistic:

$$Z = \frac{\hat{P} - P_0}{\sqrt{\frac{P_0(1-P_0)}{n}}}$$

Critical value: Since the equation of the test statistic has a Z, the critical values come from the standard normal distribution. Also, since the HA is "greater than," the CV = 1.645.

When to reject the H0: Reject the H0 when TS > 1.645.

Collected information: The results given are Yes (f = 38), No (f = 12). We need the other pieces in the test statistic. We know the sample size is 50, so $n = 50$:

$$\hat{P} = \frac{\text{frequency of outcome of interest}}{n} = \frac{\text{frequency of Yes}}{n} = \frac{38}{50} = 0.76$$

Interpret $\hat{\mathbf{P}}$: I predict that 76% of students are in favor of creating a new parking lot on campus.

Calculate the TS:

$$TS = \frac{\hat{P} - P_0}{\sqrt{\frac{P_0(1-P_0)}{n}}} = \frac{0.76 - 0.5}{\sqrt{\frac{0.5(1-0.5)}{50}}} = 3.677$$

Find the p-value: The HA is "greater than," thus the probability statement for the p-value is

$$\begin{aligned} \text{p-value } &= P(Z > TS) = P(Z > 3.68) \text{ rounded to match the tables in Appendix B} \\ &= 0.00012 \text{ See Section 9.2.3 for a review} \end{aligned}$$

Statistical decision:

Using the CV: Reject the H0, because 3.677(TS) > 1.645(CV).

Using the p-value: Reject the H0, because 0.00012(p-value)< 0.05(α).

Conclusion/interpretation: I conclude that the proportion of students who are in favor of the new parking lot is greater than the proportion of students who are not in favor of the new parking lot. Specifically, I predict that 76% of students favor the creation of a new campus parking lot.

Note: This example has a lot of detail so that each piece could be explained. The goal is to demonstrate that the interpretation is the answers to the question of interest. In practice the p-value and the CV would not both be used but are shown in the examples for illustrative purposes. The rest of the examples in this section will be more straightforward in presentation.

2. Keith Beck and Pat Gray are running for student body president (fictitious). Who will win the election for student body president: Keith Beck or Pat Gray? To answer this question, 100 students were asked, "Who are you going to vote for: Pat Gray or Keith Beck?" Results: Keith (f = 45) and Pat (f = 55). $\alpha = 0.05$.

   The question of interest: Who will win the election for student body president: Keith Beck or Pat Gray?

   Test situation: One sample proportion

   - One sample: 100 students
   - One variable: Who will you vote for?

- Categorical with two outcomes

Hypotheses: Neither outcome is in the context of the question of interest, so we pick from the question of interest the first outcome, Keith, to be the outcome of interest. There is no reference group or goal, so $P_0 = 0.5$ for this scenario.

H0: The proportion of students who will vote for Keith is equal to the proportion of students who will vote for Pat. H0: $P = 0.5$

HA: The proportion of students who will vote for Keith is *not* equal to the proportion of students who will vote for Pat. HA: $P \neq 0.5$

The HA is "unequal to" because the question of interest doesn't indicate one outcome greater or smaller than the other outcome.

Find the CV: Since the HA is an unequal to, the CV is -1.96 and 1.96.

Reject the H0 for $TS < -1.96$ or $TS > 1.96$.

Calculate the TS: $n = 100$, $\hat{P} = \frac{45}{100} = 0.45$

$$TS = \frac{\hat{P} - P_0}{\sqrt{\frac{P_0(1-P_0)}{n}}} = \frac{0.45 - 0.5}{\sqrt{\frac{0.5(1-0.5)}{100}}} = -1$$

Calculate the p-value: The HA is a "unequal to," thus the probability statement is $2P(Z > |TS|)$. This statement might seem a little odd. Figure 12.1 visually represents the probability statement. Recall, the p-value is the probability of being more extreme than the TS. In this case, that is being less than the negative value of the TS and being greater than the positive value of the TS.

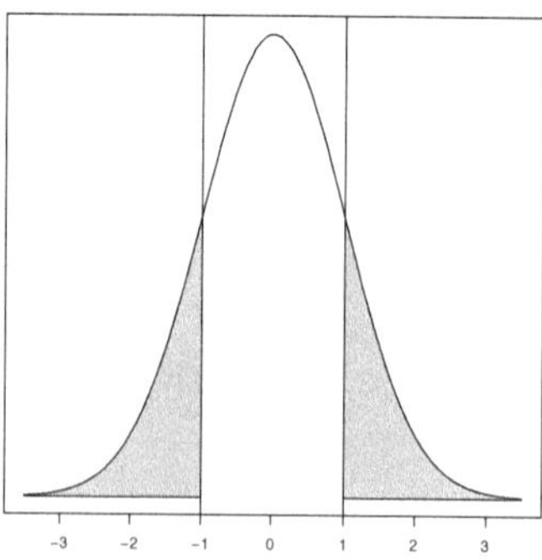

Figure 12.1: Standard normal distribution. Shaded area corresponds to the probability statement $2P(Z > |TS|)$.

The symmetry of the normal distribution allows the calculation of one of the shaded areas of Figure 12.1; then multiply by 2:

$$\begin{aligned}
\text{p-value} &= 2P(Z > |TS|) = 2P(Z > |-1|) \\
&= 2P(Z > 1) \\
&= 2(1 - P(Z < 1)) \\
&= 2(1 - 0.84134) \\
&= 2(0.15866) \\
&= 0.31732
\end{aligned}$$

Statistical decision:

Using the CV: Fail to reject the H0, because -1(TS) is not greater 1.96(CV) and -1(TS) is not smaller than -1.96(CV). The best way to think about this is with a number line:

$$\begin{array}{cccc} -1.96 & -1 & 0 & 1.96 \\ \text{CV} & \text{TS} & & \text{CV} \end{array}$$

Any value of the TS between -1.96 and 1.96 is where we fail to reject the H0. Any value outside of that range, we reject H0.

Using the p-value: Fail to reject the H0, because 0.31732(p-value) $> 0.05(\alpha)$.

Conclusion/interpretation: I conclude that the proportion of students who will vote for Keith is equal to the proportion of students who will vote for Pat. Thus, we cannot determine who will win the student body election.

3. In the second quarter of 2020, Apple's cell phone market share is 46% in the United States. At that time they planned to release the iPhone 12 during the fourth quarter of 2020. Apple (fictitious) conducted a study during the second quarter of 2020. They asked 50 American adults, "Will you buy the new iPhone 12 when it is released: Yes or No?" Apple wants to know if the iPhone 12 will increase their market share. Results: Yes (f = 35), No (f = 15).

Question of interest: Will the iPhone 12 increase Apple's cell phone market share?

Test situation: One sample proportion

- One sample: 50 American adults
- One variable: Will you buy the iPhone 12?
- Categorical with two outcomes

Hypotheses: The outcome Yes (will buy an iPhone 12) is in the context of the question of interest. Specifically, we want to compare the Yes responses to the known market share from the second quarter of 2020. Thus, $P_0 = 0.46$.

H0: The proportion of American adults who will buy the iPhone 12 is equal to Apple's cell phone market share in the second quarter of 2020. H0: $P = 0.46$.

HA: The proportion of American adults who will buy the iPhone 12 is greater than Apple's cell phone market share in the second quarter of 2020. H0: $P > 0.46$.

Find the CV: The CV is 1.645, since the HA is >.

Calculate the TS: $n = 50$, $\hat{P} = \frac{35}{50} = 0.7$

$$TS = \frac{\hat{P} - P_0}{\sqrt{\frac{P_0(1-P_0)}{n}}} = \frac{0.7 - 0.46}{\sqrt{\frac{0.46(1-0.46)}{50}}} = 3.405$$

Calculate the p-value: The probability statement is $P(Z > TS)$, because the HA is >.

$$\text{p-value } = P(Z > TS) = P(Z > 3.41) = 1 - P(Z < 3.41) = 1 - 0.99968 = 0.00032$$

Statistical decision:

Using the CV: Reject the H0, because 3.405(TS) > 1.645(CV).

Using the p-value: Reject the H0, because 0.00032(p-value)< 0.05(α).

Conclusion/interpretation: I conclude the proportion of American adults who will buy the iPhone 12 is greater than Apple's cell phone market share in the second quarter of 2020. Specifically, I predict that Apple's cell phone market share in the fourth quarter of 2020 will be 70%.

4. Does drinking warm milk before bed reduce sleepless nights? Suppose one person on 40 randomly selected mornings after drinking warm milk the night before was asked, "How did you sleep: Well or Terrible?" Results: Well (f = 30), Terrible (f = 10). $\alpha = 0.05$.

Question of interest: Does drinking warm milk before bed reduce sleepless nights?

Test situation: One sample proportion

- One sample: 40 nights
- One variable: How did you sleep?
- Categorical with two outcomes

Hypotheses: Terrible is the outcome of interest. It's in the context of the question of interest.

$P_0 = 0.5$; there is no reference group or goal.

H0: After drinking warm milk, 50% of nights are terrible nights of sleep. H0: $P = 0.5$

HA: After drinking warm milk, less than 50% of nights are terrible nights of sleep. HA: $P < 0.5$

Find the CV: CV = -1.645, the HA is "less than."

Calculate the TS: $n = 40$, $\hat{P} = 0.25$

$$TS = \frac{\hat{P} - P_0}{\sqrt{\frac{P_0(1-P_0)}{n}}} = \frac{0.25 - 0.5}{\sqrt{\frac{0.5(1-0.5)}{40}}} = -3.162$$

Calculate the p-value: The probability statement is $P(Z < TS)$, because the HA is <.

$$\text{p-value } = P(Z > TS) = P(Z < -3.16) = 0.00079$$

Statistical decision:

Using the CV: Reject the H0, because -3.162(TS) < -1.645(CV).

Using the p-value: Reject the H0, because 0.00079(p-value)$<$ 0.05(α).

Conclusion/interpretation: I conclude that less than 50% of nights are terrible nights of sleep after drinking a glass of warm milk. Specifically, I expect only 25% of nights to be terrible after drinking a glass of warm milk.

12.4 Examples With R

Since software is being used, less rounding has to occur, so the solutions here will be slightly different from those previously shown.

1. Return to the campus parking scenario (page 133):

```
> prop.test(x=38,n=50,p=0.5,alternative='greater',correct=FALSE)

1-sample proportions test without continuity
correction


data:  38 out of 50, null probability 0.5
X-squared = 13.52, df = 1, p-value = 0.000118
alternative hypothesis: true p is greater than 0.5
95 percent confidence interval:
 0.6489738 1.0000000
sample estimates:
   p
0.76
```

The TS $= \sqrt{13.52}$, the p-value $= 0.000118$, and $\hat{P} = 0.76$. Ignore the confidence interval presented. It is a more complicated version of the confidence for a proportion from that presented in Section 10.2.

2. Return to the student body president scenario (page 135):

```
> prop.test(x=45,n=100,p=0.5,alternative='two.sided',correct=FALSE)

1-sample proportions test without continuity
correction


data:  45 out of 100, null probability 0.5
X-squared = 1, df = 1, p-value = 0.3173
alternative hypothesis: true p is not equal to 0.5
95 percent confidence interval:
 0.3561454 0.5475540
sample estimates:
   p
0.45
```

The TS $= \sqrt{1}$, the p-value $= 0.3173$, and $\hat{P} = 0.45$. Ignore the confidence interval presented. It is a more complicated version of the confidence for a proportion from that presented in Section 10.2.

3. Return to the Apple market share scenario (page 137):

```
> prop.test(x=35,n=50,p=0.46,alternative='greater',correct=FALSE)

1-sample proportions test without continuity

correction


data:  35 out of 50, null probability 0.46

X-squared = 11.594, df = 1, p-value = 0.0003308

alternative hypothesis: true p is greater than 0.46

95 percent confidence interval:

 0.5854004 1.0000000

sample estimates:

  p

0.7
```

The TS $= \sqrt{11.594}$, the p-value $= 0.0003308$, and $\hat{P} = 0.7$. Ignore the confidence interval presented. It is a more complicated version of the confidence for a proportion from that presented in Section 10.2.

4. Return to the sleeping scenario (page 138):

```
> prop.test(x=10,n=40,p=0.5,alternative='less',correct=FALSE)

1-sample proportions test without continuity

correction


data:  10 out of 40, null probability 0.5

X-squared = 10, df = 1, p-value = 0.0007827

alternative hypothesis: true p is less than 0.5

95 percent confidence interval:

 0.0000000 0.3759729

sample estimates:

   p

0.25
```

The TS $= \sqrt{10}$, the p-value $= 0.0007827$, and $\hat{P} = 0.25$. Ignore the confidence interval presented. It is a more complicated version of the confidence for a proportion from that presented in Section 10.2.

Chapter 13

Multinomial

The multinomial test situation has one sample and one categorical variable, and there are more than two outcomes. The purpose of this test situation is to compare the outcomes to each other. Specifically, are the outcomes equally likely to occur or not?

13.1 Terms

Multinomial test: One sample and one categorical variable with three or more outcomes.

O: Observed frequency.

E: Expected frequency.

$\tilde{\mathbf{P}}$: Fair probability (Section 7.5.1).

$\boldsymbol{n}$: The sample size.

δ: Delta, the degrees of freedom.

$\boldsymbol{\chi^2}$: The chi-squared distribution (Section 9.3.2).

13.2 Template

Once it has been determined that the test situation is multinomial, this is how to conduct the test:

H0: All the outcomes are equally likely. H0: $P = \tilde{P}$.

HA: The outcome are not equally likely. In symbols, HA: not H0.

Equation of the test statistic:

$$\chi^2_\delta = \sum \frac{(O-E)^2}{E}$$

Null distribution: χ^2 distribution (Section 9.3.2)

Critical values:

1. δ = degrees of freedom = # outcomes -1.
2. Use the 0.05 column (since $\alpha = 0.05$) in Table A.1 to find the CV from the χ^2 distribution.

Test statistic:

1. Calculate E = expected frequency = $n\tilde{P}$ = sample size times the fair probability
2. Calculate $(O-E)$
3. Calculate $(O-E)^2$
4. Calculate $(O-E)^2/E$
5. TS = sum $(O-E)^2/E$

p-value: The probability statement is p-value = $P(\chi^2_\delta > \text{TS})$.

Statistical decision:

- Fail to reject the H0 if TS < CV or if p-value > α.
- Reject the H0 if TS > CV or if p-value < α.

Conclusion/interpretation:

- If the statistical decision is fail to reject the H0, then conclude the H0 in English.
- If the statistical decision is to reject the H0, then conclude the HA in English, and interpret largest and smallest (O-E) values.

13.3 Examples

1. Are basketball shoes equally or unequally preferred by NBA players? To test this claim, suppose 75 NBA players were asked, "Which brand of basketball shoe do you prefer: Adidas (A), Nike (N), Reebok (R)?" Results: A (f = 32), N (f = 21), R (f = 22). $\alpha = 0.05$.

Question of interest: Are basketball shoes equally or unequally preferred by NBA players?

Test situation: Multinomial

- One sample: 75 NBA players
- One variable: Which brand of basketball shoe do you prefer?
- Categorical with more than two outcomes

H0: All three brands of shoes are equally favored by NBA players.

$\tilde{P}$ = 1/(number of outcomes) = 1/3

H0: $P = 1/3$

HA: All three brands of shoes are not equally favored by NBA players. HA: not H0.

Equation of the test statistic: The O is the observed frequency. In Chapter 5 we called these frequencies f; now we are just changing the letter to O. The equation of the TS is

$$\chi^2_\delta = sum\frac{(O-E)^2}{E}$$

CV: Since the equation of the TS has the χ^2 (the null distribution), which is where the CV comes from, we first need to find the degrees of freedom. δ = number of outcomes - 1 = 3 - 1 = 2. Now that we have the degrees of freedom, we go to Table A.1. The left side is the degrees of freedom, so we go down to the row where the 2 is, since $\delta = 2$. There are four columns other than the degree of freedom columns that have headings: 0.1, 0.05, 0.025, 0.01. These are the alpha values; since $\alpha = 0.05$, we will use the 0.05 column. When we go down to 2 degrees of freedom and over to the 0.05 column, the number is 5.991. Thus CV = 5.991; reject the H0 when the TS > 5.991.

Calculate TS: We have the O or observed frequencies. We need the expected frequency, E:

$E = \tilde{P}n = (1/3)75 = 25$

The $\tilde{P}$ comes from the H0, the fair probability.

	O	E
Adidas	32	25
Nike	21	25
Reebok	22	25

I observed 32 NBA players who preferred Adidas basketball shoes. Under the assumption of fair probability I expect 25 NBA players to prefer Adidas. E is the frequency we should see if all the outcomes are equally likely. The O is the frequency we observe. The following table is how to calculate the TS:

	O	E	$(O-E)$	$(O-E)^2$	$(O-E)^2/E$
Adidas	32	25	32-25=7	$(7)^2=49$	$49/25=1.96$
Nike	21	25	21-25=-4	$(-4)^2=16$	$16/25=.64$
Reebok	22	25	22-25=-3	$(-3)^2=9$	$9/25=.36$

Then the sum of the last column is the test statistic:

$$TS = sum\frac{(O-E)^2}{E} = 1.96 + 0.64 + 0.36 = 2.96$$

p-value: The probability statement is p-value = $P(\chi^2_\delta > \text{TS}) = P(\chi^2_2 > 2.96)$.

Statistical decision: We need to compare the TS to the CV.

Fail to reject H0 because 2.96 (TS) is not greater than 5.99 (CV).

Conclusion/interpretation: I conclude all three brands of shoes are equally favored by NBA players.

2. There is an old saying, "An apple a day keeps the doctor away." Which fruit is best to eat every day? To answer this question, suppose 200 medical doctors were asked, "Which fruit is best to eat every day: Apple, Banana, Orange, Pineapple?" $\alpha = 0.05$. Results: Apple (f = 20), Banana (f = 70), Orange (f = 85), Pineapple (f = 25).

Question of interest: Which fruit is best to eat every day?

Test situation: Multinomial

- One sample: 200 doctors
- One variable: Which fruit is best to eat every day?
- Categorical with more than two outcomes

H0: The four fruits are equally favored by doctors to eat every day.

HA: The four fruits are not equally favored by doctors to eat every day.

Find the CV: CV = 7.815; δ = number of outcomes -1 = 4-1 = 3; use Table A.1

Calculate the TS: $E = \tilde{P}n = (1/4)200 = 50$

	O	E	$(O-E)$	$(O-E)^2$	$(O-E)^2/E$
Apple	20	50	-30	900	18
Banana	70	50	20	400	8
Orange	85	50	35	1225	24.5
Pineapple	25	50	-25	625	12.5

TS = 18 + 8 + 24.5 + 12.5 = 63

p-value: The probability statement is p-value = $P(\chi^2_\delta > \text{TS}) = P(\chi^2_3 > 2.96)$.

Statistical decision: Reject the H0, since the 63(TS) > 7.815(CV).

Conclusion/interpretation: I conclude that the four fruits are not equally considered best by doctors to eat every day. Specifically, oranges are considered best by doctors to eat every day (largest O - E), and apples are not considered best by doctors to eat every day (smallest O - E).

13.4 Examples With R

1. Return to the basketball shoes scenario (page 144):

```
> shoes <- c(32, 21, 22)

> chisq.test(shoes, correct=FALSE)

Chi-squared test for given probabilities


data:  shoes

X-squared = 2.96, df = 2, p-value = 0.2276
```

The "X-squared" is the test statistic. The "df" is the degrees of freedom or δ. The function also provides the p-value. In this case, the p-value is greater than 0.05, thus the statistical decision is fail to reject the null hypothesis. The same conclusion is found when using the critical value.

2. Return to the apple a day keeps the doctor away scenario (page 146):

```
> doc <- c(20, 70, 85, 25)

> chisq.test(doc, correct=FALSE)

Chi-squared test for given probabilities


data:  doc

X-squared = 63, df = 3, p-value = 1.343e-13
```

The p-value is presented in scientific notation (Section 4.5.4). This means the p-value is really small and thus less that 0.05. The statistical decision is to reject the null hypothesis, the same as using the critical value.

Chapter 14

Pre-Post (Change)

The pre-post or change test situation has one sample, one ratio variable, and an intervention. The purpose of the pre-post test is to assess the intervention. The unique part is the variable is measured twice: before the intervention and after the intervention.

14.1 Terms

Pre-post or change test: One sample, one ratio variable, and an intervention.

Intervention: An act, or program intended to change some variable's response in a predetermined direction.

Pre scores: The variable's scores on the subjects before the intervention.

Post scores: The variable's scores on the subjects after the intervention.

Change scores: The scores created from post-scores minus pre-scores (change = post - pre); in this text the change scores will always be post minus pre.

μ_C: The population mean of the change scores.

$\hat{\mu}_C$: The sample mean of the change scores.

$\mathbf{S_C}$: The standard deviation of the change scores.

$\mathbf{S_C^2}$: The variance of the change scores.

δ: Delta, the degrees of freedom.

t: The t distribution (Section 9.3.1).

The change (pre-post) test situation tries to determine if the intervention worked. Measuring a subject before and after an intervention allows us to see if the change is due to the intervention.

14.2 Template

Once it has been determined that the test situation is change (pre-post), this is how to conduct the test:

H0: The intervention results in no change on the variable in the population. H0: $\mu_C = 0$.

Alternative hypothesis: One of the following based on the question of interest:

- HA: The intervention results in an increase on the variable in the population. HA: $\mu_C > 0$.
- HA: The intervention results in a decrease on the variable in the population. HA: $\mu_C < 0$.

Equation of the test statistic:

$$t_\delta = \frac{\hat{\mu}_C}{S_C/\sqrt{n}}$$

Null distribution: t distribution (Section 9.3.1)

Critical value:

Critical value(s):

1. Find δ = degrees of freedom $= n - 1$.
2. Use Table A.2 to find the CV from the t distribution.
    - If HA: $\mu_C > 0$, use the 0.05 column and leave CV positive.
    - If HA: $\mu_C < 0$, use the 0.05 column and make CV negative.

p-value: The probability statement for the p-value depends on the HA:

- If HA: $\mu_C > 0$, p-value $= P(t_\delta > \text{TS})$
- If HA: $\mu_C < 0$, p-value $= P(t_\delta < \text{TS})$

Statistical decision:

- Fail to reject the H0, if p-value $> \alpha$ or
    - if TS $<$ CV and HA: $\mu_C > 0$

- if TS > CV and HA: $\mu_C < 0$

- Reject the H0, if p-value $< \alpha$ or
  - if TS > CV and HA: $\mu_C > 0$
  - if TS < CV and HA: $\mu_C < 0$

Conclusion/interpretation:

- If the statistical decision is to fail to reject the H0, then conclude the H0 in English.
- If the statistical decision is to reject the H0, then conclude the HA in English and interpret the estimated mean of the change ($\hat{\mu}_C$).

14.3 Examples

1. Does the diet pill Slimvox (fictitious) help women lose weight? To answer this claim, a researcher weighed seven women, then gave them Slimvox for 3 weeks. After the 3 weeks the researcher measured their weight: $\alpha = 0.05$.

	Woman 1	Woman 2	Woman 3	Woman 4	Woman 5	Woman 6	Woman 7
Pre-weight	163	157	170	162	176	213	222
Post-weight	156	152	172	162	171	207	213

Before we get into the test situation we need to talk about how the enumeration is processed and what it means. Consider the following questions:

(a) How much did the second woman weigh before the intervention?

Answer: She weighed 157 pounds.

(b) How much did the second woman weigh after the intervention?

Answer: She weighed 152 pounds.

(c) How much weight did the second woman lose using Slimvox?

Answer: She lost 5 pounds. Post - pre weight = 152 - 157 = -5 pounds.

(d) Calculate the change scores.

Answer:

(e) Interpret the sixth woman's change score.

Women	Post	Pre	Post-Pre	Change
1	156	163	$156 - 163 =$	-7
2	152	157	$152 - 157 =$	-5
3	172	170	$172 - 170 =$	2
4	162	162	$162 - 162 =$	0
5	171	176	$171 - 176 =$	-5
6	207	213	$207 - 213 =$	-6
7	213	222	$213 - 222 =$	-9

Answer: Her change score is -6. The negative means she lost weight; the 6 is 6 pounds. The interpretation is she lost 6 pounds taking Slimvox for 3 weeks.

(f) Interpret the fourth woman's change score.

Answer: She lost zero pounds after taking Slimvox for 3 weeks.

(g) Interpret the third woman's change score.

Answer: She gained 2 pounds after taking Slimvox for 3 weeks.

(h) Calculate $\hat{\mu}_C$.

Answer:

$$\hat{\mu}_C = \frac{\text{sum of change scores}}{n} = \frac{(-7) + (-5) + 2 + 0 + (-5) + (-6) + (-9)}{7} = \frac{-30}{7} = -4.29$$

(i) Interpret $\hat{\mu}_C$.

Answer: I expect the typical woman to lose 4.29 pounds after taking Slimvox for 3 weeks.

Since the mean change comes from a sample, it is not perfect. Recall from Section 8.2.2, the way we quantified the uncertainty in the sample from a ratio variable is with the standard deviation. The variance and then the standard deviation are calculated just as before. We take the change scores as a single sample and calculate the variance. This is a good opportunity to practice calculating a variance:

$$S_C^2 = 15.24$$

S^2 is the variance, so S_C^2 is the variance of the change scores. The subscripts tell us what the variance is of. Remember that to find the standard deviation is to take the square root of the variance:

$$S_C = \sqrt{S_C^2} = \sqrt{15.24} = 3.90$$

Question of interest: Does the diet pill Slimvox help women lose weight?

Intervention: Taking the Slimvox for 3 weeks.

Test situation: Change or pre-post

- 1 sample: Seven women
- 1 variable: Weight measured before and after taking Slimvox
- Ratio

H0: Women do not lose weight taking Slimvox for 3 weeks. H0: $\mu_C = 0$.

HA: Women lose weight taking Slimvox for 3 weeks. HA: $\mu_C < 0$. The decrease comes from "lose weight" in the question of interest.

Equation of test statistic:

$$t_\delta = \frac{\hat{\mu}_C}{S_C/\sqrt{n}}$$

CV: Because of the t_δ the CV comes from Table A.2. Find the degrees of freedom and use the t-Table:

(a) $\delta = n - 1 = 7 - 1 = 6$.

(b) Then the value in the 0.05 column is 1.943. Since the HA is "less than," the CV is negative.

The CV = -1.943; reject the H0 if TS < -1.943.

Calculate the TS:

$$\text{TS} = \frac{\hat{\mu}_C}{S_C/\sqrt{n}} = \frac{-4.29}{3.90/\sqrt{7}} = -2.91$$

p-value: The probability statement is p-value = $P(t_\delta < \text{TS}) = P(t_6 < -2.91)$.

Statistical decision: Reject the H0, because -2.91(TS) < -1.943(CV).

Conclusion/interpretation: I conclude that women lose weight taking Slimvox for 3 weeks. Specifically, I predict a woman to lose 4.29 pounds after taking Slimvox for 3 weeks.

2. There is a lot of talk about public schools versus charter schools. Does the movie *Waiting for Superman* increase parents' likelihood of sending their children to a charter school? To test this, suppose 50 parents were asked before and after watching the movie, "How likely are you to send your children to a charter school; 0 = 0%, 1 = 10%, ..., 10 = 100%?" $\alpha = 0.05$ Results: $\hat{\mu}_C = 2.7$ and $S_C = 4.6$.

   Question of interest: Does the movie *Waiting for Superman* increase parents' likelihood of sending their children to a charter school?

Test situation: Change or pre-post

- One sample: 50 parents
- One variable: How likely are you to send your children to a charter school?
- Ordinal variable treated as ratio (more than four outcomes)
- The intervention is watching the movie.

H0: Parents will not increase their likelihood of sending their children to a charter school after watching *Waiting for Superman.* H0: $\mu_C = 0$

HA: Parents will increase their likelihood of sending their children to a charter school after watching *Waiting for Superman.* H0: $\mu_C > 0$

Find the CV: CV $= 1.684$; $\delta = n - 1 = 50 - 1 = 49$. Since 49 is not in the table,the next lower value in the table is 40. Keep the CV positive because the HA is "greater than."

Calculate the TS:

$$\text{TS} = \frac{\hat{\mu}_C}{S_C/\sqrt{n}} = \frac{2.7}{4.6/\sqrt{50}} = 4.15$$

p-value: The probability statement is p-value $= P(t_\delta > \text{TS}) = P(t_{49} > 4.15)$.

Statistical decision: Reject the H0, because $4.15(TS) > 1.684(CV)$.

Conclusion/interpretation: I conclude that parents will increase their likelihood of sending their children to a charter school after watching *Waiting for Superman.* Specifically, I expect parents to increase their likelihood by 2.7 or 27%.

14.4 Examples With R

1. Return to the diet pill scenario (page 151):

```
> changeWeight <- postWeight-preWeight
> changeWeight

[1] -7 -5  2  0 -5 -6 -9

> t.test(x=changeWeight, alternative='less',mu=0)

One Sample t-test
```

```
data:  changeWeight
t = -2.9047, df = 6, p-value = 0.01358
alternative hypothesis: true mean is less than 0
95 percent confidence interval:
      -Inf -1.418703
sample estimates:
mean of x
-4.285714
```

The "t" is the test statistic value. The "df" is the degrees of freedom or δ. The p-value is less than 0.05; thus, the statistical decision is reject the H0. The "mean of x" is the mean of the change scores or $\hat{\mu}_C$.

2. Return to the public school versus charter schools scenario (page 153):

```
> meanChange <- 2.7
> sdChange <- 4.6
> nChange <- 50

> TS <- meanChange/(sdChange/sqrt(nChange))
> TS
[1] 4.150409
> pValue <- pt(TS,df=nChange-1, lower.tail=FALSE)
> pValue
[1] 6.592199e-05
```

The p-value is less than 0.05; thus, the conclusion is the same as before.

Chapter 15

Homogeneity

The homogeneity test situation has two samples and one categorical variable. The purpose of this test situation is to compare two populations. The samples are the surrogate for the populations. The categorical variable is what is used to make the comparison.

15.1 Terms

Homogeneity test: Two samples, one categorical variable.

O: Observed frequency.

E: Expected frequency.

Contingency table: A table with two or more rows and two or more columns. The boxes formed by the table are called cells. The numbers in the cells are the frequencies.

Row marginal sum: The sum of the frequencies in a particular row.

Column marginal sum: The sum of the frequencies in a particular column.

Total sum: The sum of all the frequencies; or the sum of all the row marginal sums; or the sum of all the column marginal sums.

This test situation requires the two samples to be independent. See Section 3.4 for a review of independent samples.

15.2 Contingency Table

The basics of a contingency table need to be explained before the test situation is explained. Here is a generic 2-by-3 (2 x 3) contingency table:

O11	O12	O13
O21	O22	O23

We use the letter O because they are observed frequencies. The number refers to which cell it is in. The first number is the row and the second number is the column. Then we can add the marginal sums to the table:

O11	O12	O13	R1
O21	O22	O23	R2
C1	C2	C3	T

We give each of the marginal sums a letter and number:

- R1 = row 1 marginal sum = O11 + O12 + O13
- R2 = row 2 marginal sum = O21 + O22 + O23
- C1 = column 1 marginal sum = O11 + O21
- C2 = column 2 marginal sum = O12 + O22
- C3 = column 3 marginal sum = O13 + O23
- T = total sum = O11 + O12 + O13+O21 + O22 + O23 = R1 + R2 = C1 + C2 + C3

Now these marginal sums can be generalized to any size table. Why are they called marginal sums? Because the sums are put in the margins of the table. Since homogeneity compares two samples, we put the samples as the rows and the outcomes to the variable as the columns.

15.3 Template

Once it has been determined that the test situation is homogeneity, here is how to conduct the test:

H0: The first population's preference to the outcomes is the same as the second population's preference to the outcomes.

HA: The first population's preference to the outcomes is different from the second population's preference to the outcomes.

Equation of the statistic:

$$\chi^2_\delta = sum\frac{(O-E)^2}{E}$$

Null distribution: χ^2 distribution (Section 9.3.2).

Critical value:

1. δ = degrees of freedom = (# row -1)(# column -1).
2. Use the 0.05 column (since $\alpha = 0.05$) in Table A.1 to find the CV from the χ^2 distribution.

Calculate the TS: The O is the table of observed frequencies. The E is the table of expected frequencies.

This is calculated using the marginal sums from the O table. Here is an example:

$E11 = \frac{R1*C1}{T}$	$E12 = \frac{R1*C2}{T}$	$E13 = \frac{R1*C3}{T}$	R1
$E21 = \frac{R2*C1}{T}$	$E22 = \frac{R2*C2}{T}$	$E23 = \frac{R2*C3}{T}$	R2
C1	C2	C3	T

1. Calculate $O - E$ table
2. Calculate $(O-E)^2$ table
3. Calculate $(O-E)^2/E$ table
4. TS = sum $(O-E)^2/E$ table

p-value: The probability statement is p-value = $P(\chi^2_\delta > \text{TS})$.

Statistical decision:

- Fail to reject the H0 if TS < CV or if p-value > α.
- Reject the H0 if TS > CV or if p-value < α.

Conclusion/interpretation:

- If the statistical decision is to fail to reject the H0, then conclude the H0 in English.
- If the statistical decision is to reject the the H0, then conclude the HA in English and interpret the two largest O - E values.

15.4 Examples

1. Do women and men enjoy reading the same genre of book? Suppose 50 women and 50 men were asked, "What is your favorite genre of book to read: Fantasy, Romance, Western?" $\alpha = 0.05$. $\alpha = 0.05$

   Results:

	Fantasy	Romance	Western
Women	10	35	5
Men	30	5	15

   This is a 2 x 3 (two-by-three) contingency table. The samples are the rows, and the columns are the outcomes to the variable. Before we get into the hypothesis testing, consider a few question about the table:

   (a) What value is in the O21 cell?

   Answer: The second row, the sample of men, and the first column, the outcome Fantasy. O21 = 30.

   (b) Interpret the O21 cell.

   Answer: Thirty of the men measured reported their favorite book genre is Fantasy.

   (c) Interpret the O13 cell.

   Answer: Five of the women measured reported their favorite book genre is Western.

   Now to get into the test situation:

   Question of interest: Do women and men enjoy reading the same genre of book?

   Test situation: Homogeneity

   - Two samples: 50 women and 50 men
   - One variable: Favorite genre of book to read
   - Categorical

   H0: Women's book genre preferences are the same as men's book genre preferences.

   HA: Women's book genre preferences are different from men's book genre preferences.

   Equation of test statistic:

$$\chi_{\delta}^2 = sum\frac{(O-E)^2}{E}$$

Critical value: The equation of the test statistic has the χ^2, so use Table A.1. δ = degrees of freedom = (# samples -1)(# outcomes -1) = (2-1)(3-1) = 2. Use the 0.05 column in Table A.1. CV = 5.991; reject the H0 if TS > 5.991.

Calculate the TS: This is done through a series of tables. But first we need the marginal sums:

O	Fantasy	Romance	Western	
Women	10	35	5	R1 = 50
Men	30	5	15	R2 = 50
	C1 = 40	C2 = 40	C3 = 20	T = 100

Notice how the row marginal sums equal the sample size. This will always be the case. To create the E table we use the marginal sums:

E	Fantasy	Romance	Western	
Women	$50*40/100 = 20$	$50*40/100 = 20$	$50*20/100 = 10$	R1 = 50
Men	$50*40/100 = 20$	$50*40/100 = 20$	$50*20/100 = 10$	R2 = 50
	C1 = 40	C2 = 40	C3 = 20	

We are not comparing the outcomes to each other but comparing women and men. Assuming the H0, or that women and men have the same preferences, the expected values are the same for each outcome. I expect 20 women and 20 men to like fantasy books. I expect 20 women and 20 men to like romance books. I expect 10 women and 10 men to like western books.

Note: The sum of the first row of the E table is also 50. The row sums and columns are the same for the E table and the O table.

Finish TS calculation: Now that we have the O and E tables, we can finish with the TS:

$O - E$	Fantasy	Romance	Western
Women	$10 - 20 = -10$	$35 - 20 = 15$	$5 - 10 = -5$
Men	$30 - 20 = 10$	$5 - 20 = -15$	$15 - 10 = 5$

You match up cell by cell and subtract the O table from the E table. These O - E values are deviations; notice how for the O - E table all the marginal sums equal zero. This provides a good check to make sure the calculations are right:

$(O - E)^2$	Fantasy	Romance	Western
Women	$(-10)^2 = 100$	$(15)^2 = 225$	$(-5)^2 = 25$
Men	100	225	25

$(O-E)^2/E$	Fantasy	Romance	Western
Women	$100/20 = 5$	$225/20 = 11.25$	$25/10 = 2.5$
Men	5	11.25	2.5

Then the sum of this last table is the value of the TS:

TS $= 5 + 11.25 + 2.5 + 5 + 11.25 + 2.5 = 37.5$

p-value: The probability statement is p-value $= P(\chi^2_\delta > \text{TS}) = P(\chi^2_2 > 37.5)$.

Statistical decision: Reject the H0, because 37.5(TS) $>$ 5.991(CV).

Conclusion/interpretation: I conclude that women's favorite book genres are different from men's favorite book genres. Specifically, women favor romance books ($O - E = 15$, in 1,2 cell) and men prefer fantasy books ($O - E = 10$, in 2,1 cell).

2. Is it safer to go to school at UW or UW? To answer this question, suppose 50 students from the University of Wyoming and 50 students from the University of Washington were asked if they feel safe on campus (Yes or No). $\alpha = 0.05$. Results:

	Yes	No
Wyoming	40	10
Washington	20	30

Question of interest: Is it safer to go to school at UW or UW?

Test situation: Homogeneity

- Two samples: 50 Wyoming students and 50 Washington students
- One variable: Feel safe on campus
- Categorical

H0: Students at Wyoming feel the same about campus safety as do Washington students.

HA: Wyoming students feel differently about campus safety than Washington students.

Find the CV: CV $= 3.841$; $\delta = (2-1)(2-1) = 1$.

Calculate the TS: Create the tables to calculate the TS:

O	Yes	No	
Wyoming	40	10	50
Washington	20	30	50
	60	40	

E	Yes	No	
Wyoming	30	20	50
Washington	30	20	50
	60	40	

$O - E$	Yes	No
Wyoming	10	-10
Washington	-10	10

$(O - E)^2/E$	Yes	No
Wyoming	3.33	5
Washington	3.33	5

TS $= 3.33 + 5 + 3.33 + 5 = 16.66$

p-value: The probability statement is p-value $= P(\chi^2_\delta > \text{TS}) = P(\chi^2_2 > 16.66)$.

Statistical decision: Reject the H0, because 16.66(TS) > 3.841(CV).

Conclusion/interpretation: Wyoming students feel differently about campus safety than Washington students. Specifically, Wyoming students feel safe (largest O - E) and Washington students do not feel safe (second largest O - E).

15.5 Examples With R

The argument `correct=FALSE` in the `chisq.test` function ensures the calculations match what is presented in this chapter.

1. Return to the book genre scenario (page 159):

```
> Obook <- matrix(c(10,30,35,5,5,15), nrow=2, ncol=3)

> (bookResult <- chisq.test(Obook, correct=FALSE))

Pearson's Chi-squared test


data:  Obook

X-squared = 37.5, df = 2, p-value = 7.194e-09

## O-E table:

> bookResult$observed-bookResult$expected

     [,1] [,2] [,3]

[1,]  -10   15   -5

[2,]   10  -15    5
```

The "X-squared" is the test statistic value. The "df" is the degrees of freedom, or δ. The p-value is less than 0.05; thus, reject the H0 just like before.

2. Return to the safer school scenario (page 161):

```
> Oschool <- matrix(c(40,20,10,30), nrow=2, ncol=3)

> (schoolResult <- chisq.test(Oschool, correct=FALSE))

Pearson's Chi-squared test


data:  Oschool

X-squared = 16.667, df = 1, p-value = 4.456e-05

## O-E table:

> schoolResult$observed-schoolResult$expected

      [,1] [,2]

[1,]    10  -10

[2,]   -10   10
```

The p-value is less than 0.05, leading to the same conclusion.

Chapter 16

Two Independent Samples t

The two independent samples t test situation has two independent samples and one ratio variable. This test situation is similar to homogeneity in that the purpose is to compare two populations using two samples. The test situation in this chapter uses a ratio variable to make the comparison.

16.1 Terms

Two independent samples t test: Two independent samples, one ratio variable.

Estimated mean difference: The difference between the sample means. $\hat{\mu}_1 - \hat{\mu}_2$

$\mathbf{S_P^2}$: The pooled variance. The weighted average of the sample variances.

$\mathbf{S_P}$: The pooled standard deviation. The square root of the pooled variance.

This test situation requires the two samples to be independent. See Section 3.4 for a review of independent samples.

16.2 Template

Once it has been determined that the test situation is two independent samples t, here is how to conduct the test:

H0: The mean of population 1 equals the mean of population 2. H0: $\mu_1 = \mu_2$.

Alternative hypothesis: One of the following based on the question of interest:

- HA: The mean of population 1 is greater than the mean of population 2. HA: $\mu_1 > \mu_2$.
- HA: The mean of population 1 is less than the mean of population 2. HA: $\mu_1 < \mu_2$.
- HA: The mean of population 1 is different than the mean of population 2. HA: $\mu_1 \neq \mu_2$.

Equation of the test statistic:

$$t_\delta = \frac{\hat{\mu}_1 - \hat{\mu}_2}{S_P\sqrt{\frac{1}{n_1} + \frac{1}{n_2}}}$$

$$S_P^2 = \frac{(n_1 - 1)S_1^2 + (n_2 - 1)S_2^2}{n_1 + n_2 - 2}$$

$$S_P = \sqrt{S_P^2}$$

Null distribution: t distribution (Section 9.3.1)

Critical value(s):

1. Find δ = degrees of freedom $= n_1 + n_2 - 2$.
2. Use Table A.2 to find the CV from the t distribution.
   - If HA: $\mu_1 > \mu_2$, use the 0.05 column and leave CV positive.
   - If HA: $\mu_1 < \mu_2$, use the 0.05 column and make CV negative.
   - If HA: $\mu_1 \neq \mu_2$, use the 0.025 column and use both positive and negative CV.

p-value: The probability statement for the p-value depends on the HA:

- If HA: $\mu_1 > \mu_2$, p-value $= P(t_\delta > \text{TS})$.
- If HA: $\mu_1 < \mu_2$, p-value $= P(t_\delta < \text{TS})$.
- If HA: $\mu_1 \neq \mu_2$, p-value $= 2P(t_\delta > |\text{TS}|)$.

Statistical decision:

- Fail to reject the H0 if p-value $> \alpha$ or
  - if TS $<$ CV and HA: $\mu_1 > \mu_2$
  - if TS $>$ CV and HA: $\mu_1 < \mu_2$
  - if TS $>$ negative CV and TS $<$ positive CV and HA: $\mu_1 \neq \mu_2$
- Reject the H0 if p-value $< \alpha$ or

– if TS > CV and HA: $\mu_1 > \mu_2$

– if TS < CV and HA: $\mu_1 < \mu_2$

– if TS < negative CV or TS > positive CV and HA: $\mu_1 \neq \mu_2$

Conclusion/interpretation:

- If the statistical decision is to fail to reject the H0, then conclude the H0 in English.
- If the statistical decision is to reject the H0, then conclude the HA in English and interpret the estimated mean difference ($\hat{\mu}_1 - \hat{\mu}_2$).

16.3 Examples

1. Are gas prices more expensive in Fort Collins, Colorado, than Laramie, Wyoming? To answer this question, suppose the price of gas was recorded from seven randomly selected gas stations in Fort Collins and eight gas stations in Laramie. $\alpha = 0.05$. Results:

									Mean	Variance
Laramie:	3.62	3.16	2.35	3.04	2.74	2.51	3.47	2.69	2.95	0.21
Fort Collins:	3.16	4.25	4.11	3.37	3.29	3.72	3.65		3.65	0.17

Question of interest: Are gas prices more expensive in Fort Collins than Laramie?

Test situation: Two independent samples t

- Two samples: Eight gas stations in Laramie, seven gas stations in Fort Collins
- One variable: Price of gas
- Ratio

H0: The mean gas price in Fort Collins in equal to the mean gas price in Laramie. H0: $\mu_1 = \mu_2$.

HA: The mean gas price in Fort Collins is greater than the mean gas price in Laramie. HA: $\mu_1 > \mu_2$.

Equation of the TS:

$$t_\delta = \frac{\hat{\mu}_1 - \hat{\mu}_2}{S_P\sqrt{\frac{1}{n_1} + \frac{1}{n_2}}}$$

Find CV: We know from the equation of the TS that the CV comes from the Table A.2. First, we need the degrees of freedom: $\delta = n_1 + n_2 - 2 = 7 + 8 - 2 = 13$. Since the HA is "greater than," look in the 0.05 column and leave the value as positive: CV = 1.771; reject the H0 for TS > 1.771.

Calculate the TS: (a) First, we need the pooled variance:

$$S_P^2 = \frac{(n_1-1)S_1^2 + (n_2-1)S_2^2}{n_1+n_2-2} = \frac{(7-1)0.17 + (8-1)0.21}{7+8-2} = 0.19$$

(b) Now we need the pooled standard deviation:

$$S_P = \sqrt{S_P^2} = \sqrt{.19} = 0.44$$

(c) Now we have all the pieces to calculate the TS:

$$TS = \frac{\hat{\mu}_1 - \hat{\mu}_2}{S_P\sqrt{\frac{1}{n_1} + \frac{1}{n_2}}} = \frac{3.65 - 2.95}{0.44\sqrt{\frac{1}{7} + \frac{1}{8}}} = \frac{0.7}{0.23} = 3.07$$

Note: In the order of subtraction, the mean for Fort Collins came first because it is listed first in the HA. Always match the order in the HA.

p-value: The probability statement is p-value $= P(t_{13} > \text{TS}) = P(t_{13} > 3.07)$.

Statistical decision: Reject the H0 because 3.07(TS) $>$ 1.771(CV).

Conclusion/interpretation: I conclude the mean gas price in Fort Collins is greater than the mean gas price in Laramie. Specifically I expect to pay 70 cents more per gallon in Fort Collins.

2. Who drives faster on Grand Ave., men or women? Suppose with 10 men and 10 women driving on Grand Ave., speed was measured using a radar gun in miles per hour. $\alpha = 0.05$. Results:

	$\hat{\mu}$	S
Women	33	6.3
Men	29.2	5.7

Question of interest: Who drives faster on Grand Ave., men or women?

Test situation: Two independent samples t test

- Two samples: 10 women, 10 men
- One variable: Driving speed on Grand Ave.
- Ratio

H0: Women's average driving speed is the same as men's average driving speed on Grand Ave.

H0: $\mu_1 = \mu_2$

HA: Women's average driving speed is not the same as men's average driving speed on Grand Ave.

HA: $\mu_1 \neq \mu_2$

The word "faster" is not between women and men; thus, it is an unequal alternative hypothesis.

Find the CV: $\delta = n_1 + n_2 - 2 = 10 + 10 - 2 = 18$. The HA is not equal to, so use the 0.025 column, and use both the positive and negative. The CV is -2.101 and 2.101; reject the H0 if TS $>$ 2.101 or if TS $<$ -2.101.

Calculate the TS: Since we were given the standard deviations, we need to calculate the variances first:

$$S_1^2 = (5.7)^2 = 32.49 \qquad S_2^2 = (6.3)^2 = 39.69$$

Note: Always remember to plug variances into the pooled variance formula:

$$S_P^2 = \frac{(n_1 - 1)S_1^2 + (n_2 - 1)S_2^2}{n_1 + n_2 - 2} = \frac{(10 - 1)32.49 + (10 - 1)39.69}{10 + 10 - 2} = 36.09$$

Now the pooled standard deviation: $S_P = \sqrt{36.09} = 6.01$

$$TS = \frac{\hat{\mu}_1 - \hat{\mu}_2}{S_P\sqrt{\frac{1}{n_1} + \frac{1}{n_2}}} = \frac{33 - 29.2}{6.01\sqrt{\frac{1}{10} + \frac{1}{10}}} = 1.41$$

p-value: The probability statement is p-value $= 2P(t_\delta > |\text{TS}|) = 2P(t_{18} > 1.41)$.

Statistical decision: Fail to reject the H0. Since 1.41(TS) is not greater than 2.101(CV) *and* 1.41(TS) is not less than -2.101(CV), the TS is in between the critical values.

Interpretation: I conclude that women's average driving speed is the same as men's average driving speed on Grand Ave.

3. Chicago, Illinois is called the Windy City. Does Chicago have less wind than Laramie, Wyoming? To test this claim, suppose 32 randomly selected days out of the year were used to measure the wind speed (in mph) in Laramie and 31 days in Chicago. $\alpha = 0.05$ Results:

	Mean	Standard Deviation
Chicago	10.3	2.7
Laramie	12.9	3.1

Question of interest: Does Chicago have less wind than Laramie?

Test situation: Two independent samples t test

- Two samples: 31 days in Chicago and 32 days in Laramie
- One variable: Wind speed
- Ratio

H0: The mean wind speed in Chicago is the same as the mean wind speed in Laramie.

H0: $\mu_1 = \mu_2$.

HA: The mean wind speed in Chicago is less than the mean wind speed in Laramie.

HA: $\mu_1 < \mu_2$.

Find the CV: CV = -1.671; $\delta = n_1 + n_2 - 2 = 31 + 32 - 2 = 61$; but use 60. Use the 0.05 column and its negative because the HA is "less than."

Find the TS: We need the pooled standard deviation first:

Chicago variance $= S_1^2 = (2.7)^2 = 7.29$; Laramie variance $= S_2^2 = (3.1)^2 = 9.61$

$$
\begin{aligned}
S_P^2 &= \frac{(n_1 - 1)S_1^2 + (n_2 - 1)S_2^2}{n_1 + n_2 - 2} = \frac{(31 - 1)7.29 + (32 - 1)9.61}{31 + 32 - 2} = 8.47 \\
S_P &= \sqrt{8.47} = 2.91
\end{aligned}
$$

Now everthing can be put together to calculate the TS:

$$
TS = \frac{\hat{\mu}_1 - \hat{\mu}_2}{S_P\sqrt{\frac{1}{n_1} + \frac{1}{n_2}}} = \frac{10.3 - 10.3}{2.91\sqrt{\frac{1}{31} + \frac{1}{32}}} = -3.55
$$

p-value: The probability statement is p-value $= P(t_\delta < \text{TS}) = P(t_{60} < -3.55)$.

Statistical decision:: Reject the H0 because $-3.55(TS) < -1.671(CV)$.

Conclusion/interpretation: I conclude that the mean wind speed in Chicago is less than the mean wind speed in Laramie. Specifically, I expect the wind to be 2.6 mph slower in Chicago than Laramie.

16.4 Examples With R

The results are a little different from those calculated in the original example; this difference is a rounding difference.

1. Return to the gas price scenario (page 166):

```
> t.test(fortCollins, laramie, alternative='greater', var.equal=TRUE)

Two Sample t-test


data:  fortCollins and laramie

t = 3.1195, df = 13, p-value = 0.004068

alternative hypothesis: true difference in means is greater than 0

95 percent confidence interval:

 0.3036874       Inf

sample estimates:

mean of x mean of y

   3.6500    2.9475

> mean(fortCollins) - mean(laramie) # mean difference

[1] 0.7025
```

The "t" is the test statistic value. The "df" is the degrees of freedom, or δ. The p-value is less than 0.05; thus, the conclusion to reject the H0 is the same as before. The mean difference is calculated using the `mean` function and subtracting.

2. Return to the driving speed scenario (page 167):

```
> meanDiffSpeed <- meanWomen - meanMen #mean difference

> dfSpeed <- nWomen + nMen - 2 # degrees of freedom

> ## pooled variance

> Sp2Speed <- (sdWomen^2 * (nWomen-1) + sdMen^2 * (nMen-1))/dfSpeed

> ## test statistic value

> tsSpeed <- meanDiffSpeed/(sqrt(Sp2Speed)*sqrt(1/nWomen + 1/nMen))

> ## p-value

> pt(abs(tsSpeed), df=dfSpeed, lower.tail=FALSE)*2

[1] 0.1743068
```

The p-value is greater than 0.05; thus, it is the same statisticial decision of failing to reject the H0.

3. Return to the Windy City scenario (page 168):

```
> meanDiffWind <- meanChicago - meanLaramie #mean difference
> dfWind <- nChicago + nLaramie - 2 # degrees of freedom
> ## pooled variance
> Sp2Wind <- (sdChicago^2 * (nChicago-1) + sdLaramie^2 * (nLaramie-1))/dfWind
> ## test statistic value
> tsWind <- meanDiffWind/(sqrt(Sp2Wind)*sqrt(1/nChicago + 1/nLaramie))
> ## p-value
> pt(tsWind, df=dfWind)
[1] 0.0003800662
```

The p-value is less than 0.05 and we reach the same conclusion as described previously.

Chapter 17

Correlation

The correlation test situation has one sample and two ratio variables. The purpose of this test situation is to compare the variables—specifically, are the variables related to each other?

17.1 Terms

Correlation test: One sample with two ratio variables.

Scatterplot: A graphical display of two ratio variables. This is used to see the pattern between the variables.

Covariance: A numerical assignment to the pattern in the scatterplot. The covariance can be any number, but the three big distinctions are positive, negative, and zero covariance.

- Positive covariance means large values of variable 1 are associated with large values of variable 2; or small values of variable 1 are associated with small values of variable 2.
- Negative covariance means large values of variable 1 are associated with small values of variable 2; or small values of variable 1 are associated with large values of variable 2.
- Zero covariance means large values of variable 1 are associated with both large and small values of variable 2. There is no association between the variables.

Note: The covariance assesses only linear patterns.

Correlation: A scale of the covariance to be between -1 and 1. The positive, negative, and zero correlation has the same interpretation as the covariance. The correlation just scales the magnitude to between -1 and 1.

r: The population correlation value.

r̂: The sample correlation value.

17.2 Scatterplots

Think back to the wonderful days in algebra class with order pairs of X and Y. This is the same sort of thing with variable 1 and variable 2. Let's look at a few examples:

1. Two generic variables are measured on five different people. Here are the results:

Person	Variable 1	Variable 2
1	3	2
2	2	4
3	4	5
4	5	3
5	1	1

Table 17.1: Example of two different variables measured on the sample people.

We plot the values for each variable for each person. For the first person, go over to 3 on the horizontal axis and up to 2 on the vertical axis and draw a point. Do this for every person, and we have a scatterplot, as seen in Figure 17.1.

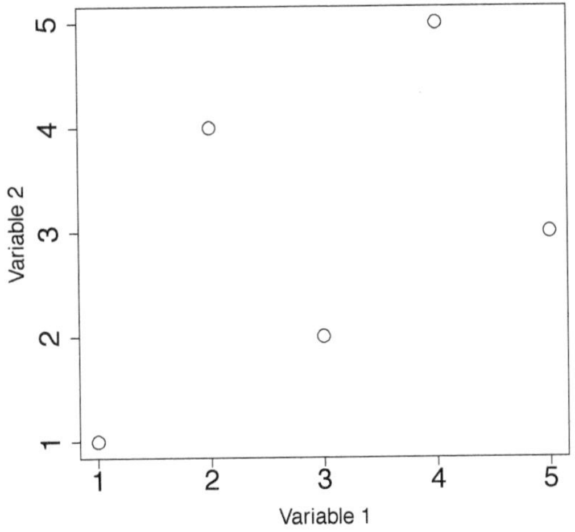

Figure 17.1: Scatterplot of generic variables.

2. Fast food is often considered unhealthy because of its fat and sodium content. Are sodium and fat related to each other? I (fictitiously) measured the fat in grams and sodium in milligrams (mg) for seven popular fast food meals. Enumeration:

Sodium:	920	1200	1010	860	1180	940	1260
Fat:	19	31	34	35	39	39	23

Put the sodium on the horizontal axis and fat on the vertical axis. The scatterplot will look as presented in Figure 17.2.

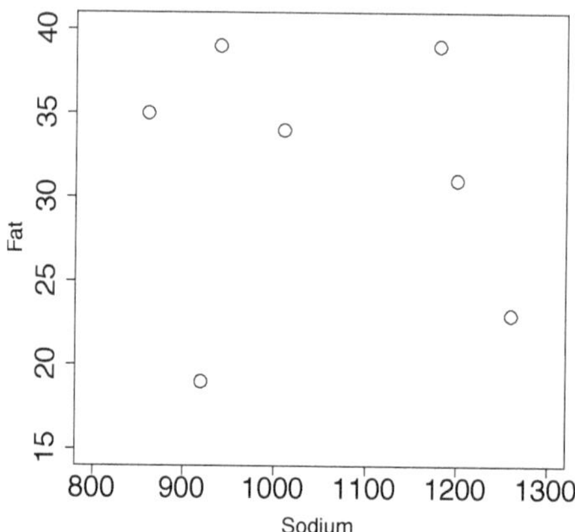

Figure 17.2: Scatterplot of sodium and fat content from popular fast food meals.

3. Is the GRE quantitative score related to GPA in graduate school? Suppose 21 students' GRE scores and their graduate GPA scores were recorded. The scatterplot is presented in Figure 17.3.

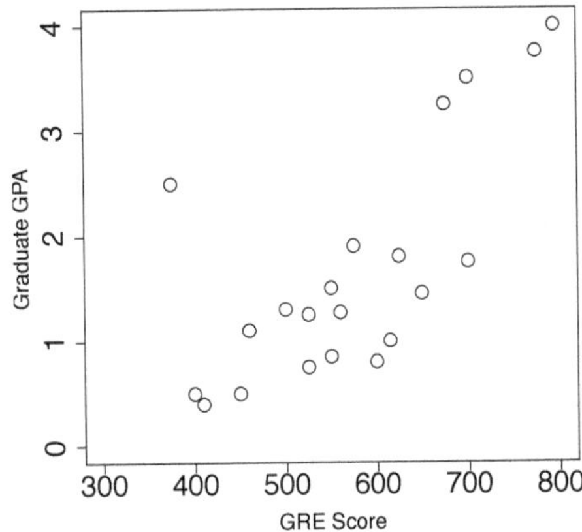

Figure 17.3: Scatterplot GRE scores and graduate school GPA scores from graduate students.

Notice how larger values of GRE score go with larger GPA, and smaller values of GRE score go with smaller values of GPA. The pattern moves from the lower left to the upper right. This is a positive correlation or relationship. The interpretation of this pattern is, as the GRE score increases, the graduate GPA also increases. Also the opposite is true: as GRE scores decrease, the graduate GPA also decreases.

Note: These statements do not apply for every person but describe the overall pattern. Some people do not follow the pattern, but most do, hence the pattern we see.

4. It has been hypothesized that the consumption of coffee in the evening before going to bed can be adversely related to your ability to sleep. To examine this claim, suppose 20 graduate students volunteered to participate in our study and were asked two questions:
   Question 1: How much coffee (measured in ounces) did you drink last night before going to bed?
   Question 2: How many hours of sleep did you get last night?

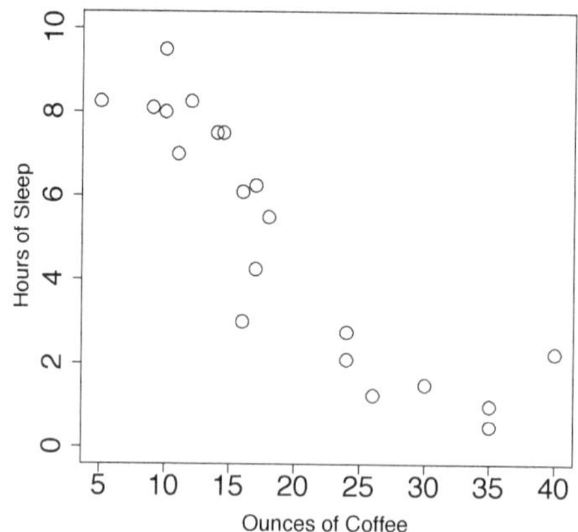

Figure 17.4: Scatterplot of coffee consumption and hours of sleep for graduate students.

The scatterplot between ounces of coffee and hours of sleep is presented in Figure 17.4. The pattern moves from the upper left corner to the lower right corner. This is a negative correlation or relationship. The interpretation is, as the amount of coffee consumed increases, the hours of sleep decreases, or, as the amount of coffee consumed decreases, the amount of sleep increases. The variables move in opposite directions.

5. Look back to Figure 17.2, the fast food example. There is no real pattern; nothing moves from one corner to another. This is a zero correlation. The interpretation is the amount of sodium in a fast food meal is not related to the amount of fat.

6. We have considered examples of positive, negative, and zero correlation. Let's see some patterns and their correlation value. Figure 17.5 presents four scatterplots with correlation values: 0.2, 0.5, 0.7, 0.95. Match up the scatterplot with the correlation value.

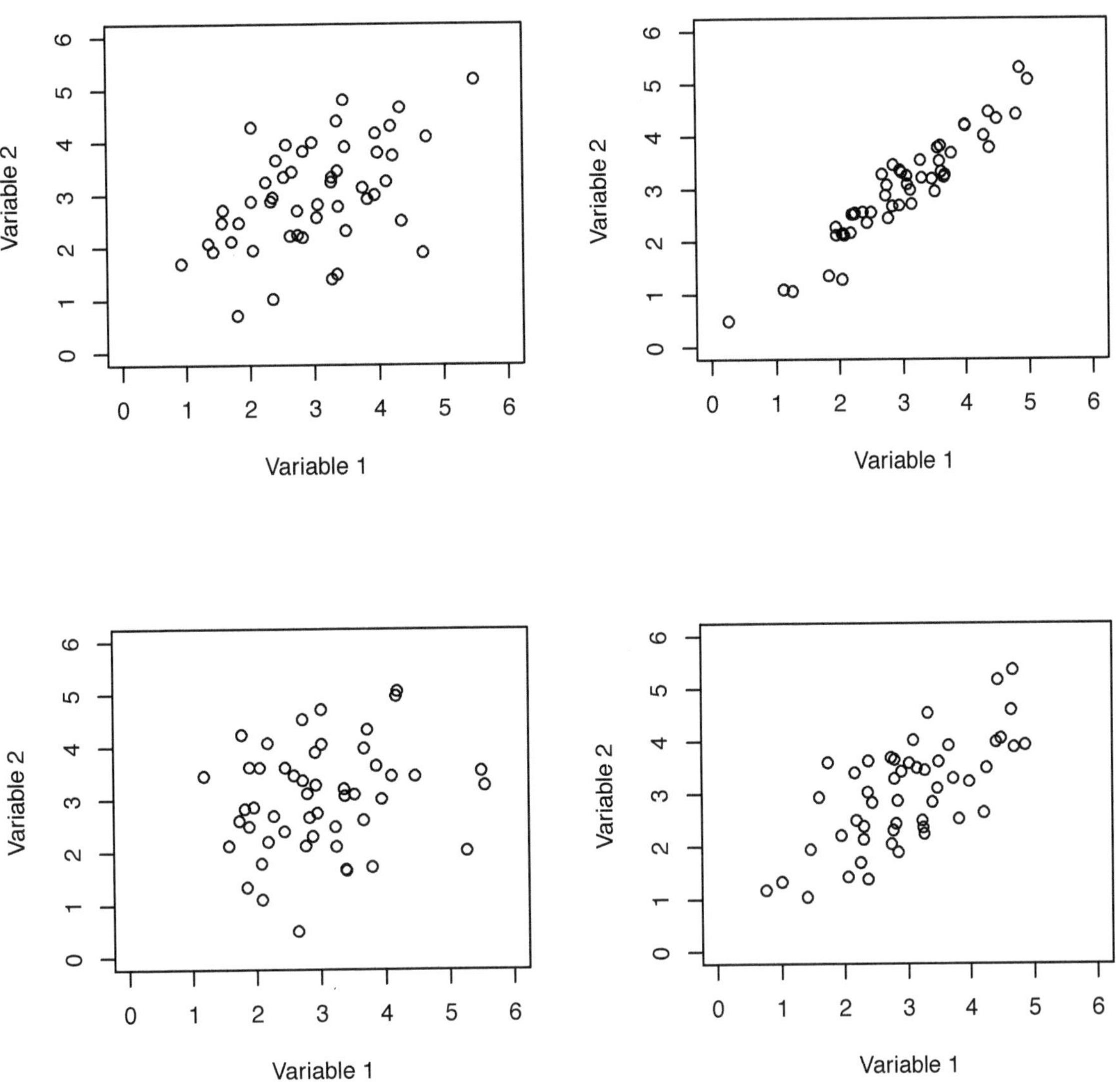

Figure 17.5: Four scatterplots depicting different correlation values.

The 0.95 is the easiest to spot. It is the scatterplot in the top right. The rest are a little hard to tell. The top left is 0.5, bottom left is 0.2, and the bottom right is 0.7. The larger the correlation value, the tighter the pattern is about a straight line.

7. Figure 17.6 presents four scatterplots. They have correlation values: 0, -0.32, -0.65, -0.82. Match up the scatterplot with the correlation value.

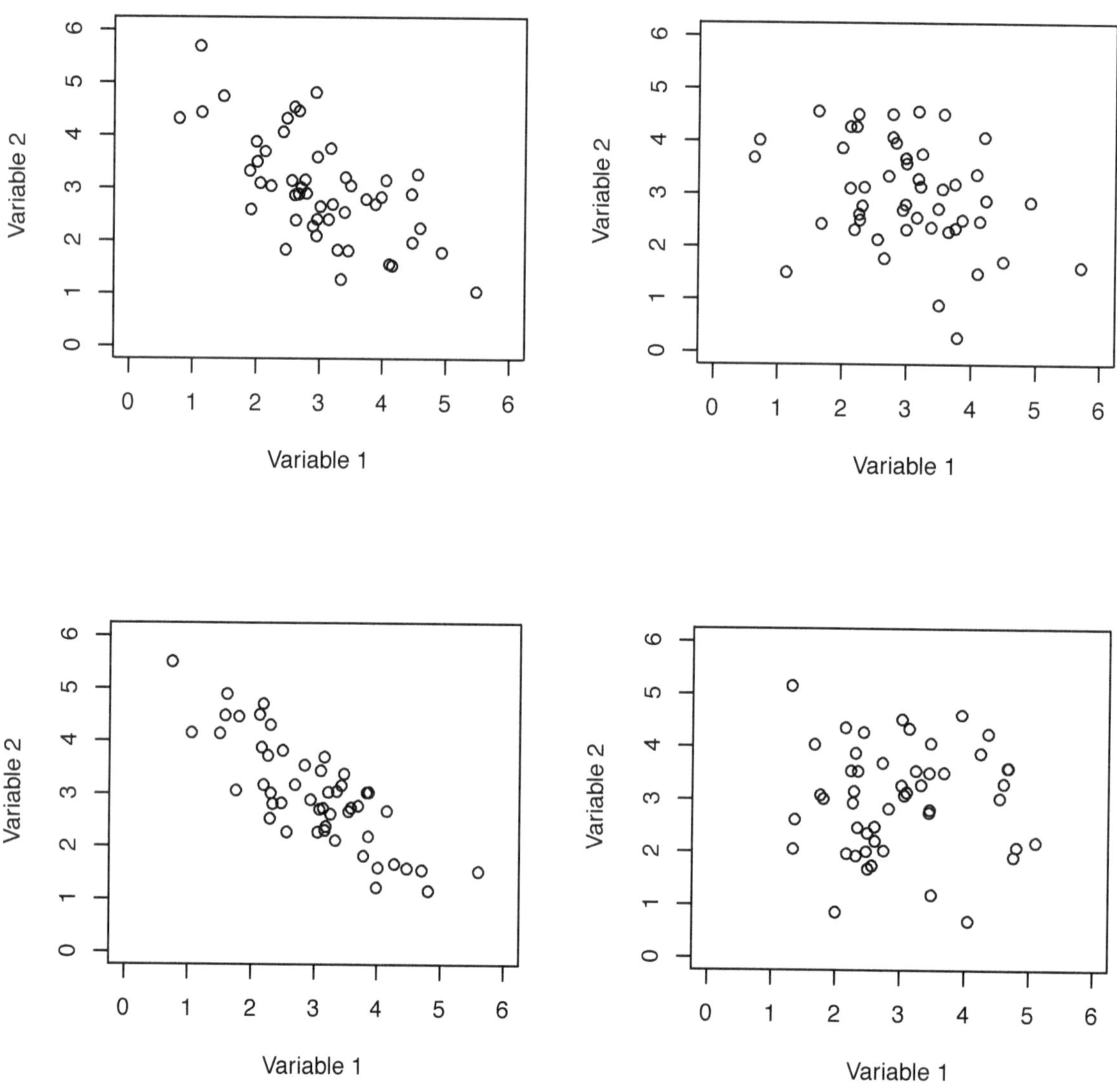

Figure 17.6: Four scatterplots depicting different correlation values.

When the correlation is close to 1 or -1, it is easier to tell. After that it gets harder. The top left is -0.65, top right is -0.32, bottom left is -0.82, and bottom right is 0.

8. What is the correlation for the pattern in Figure 17.7?

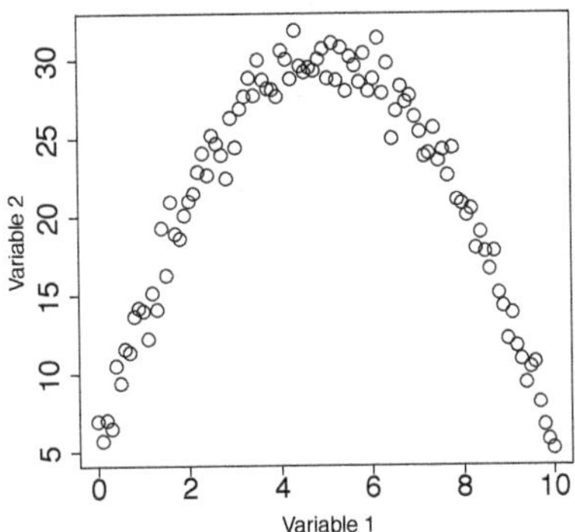

Figure 17.7: Scatterplot depicting an interesting relationship.

Even though there is a clear pattern, the correlation is zero. **Correlation assesses only linear patterns**; it does not assess nonlinear patterns.

17.3 Covariance Calculation

The equation for the covariance is

$$covariance = \frac{\sum_{i=1}^{n}(x_i - \hat{\mu}_X)(y_i - \hat{\mu}_Y)}{n-1}$$

Once you have the covariance, the equation to find the correlation is

$$\hat{r} = \frac{covariance}{\sqrt{S_X^2 S_Y^2}}$$

Let's go through an example. Here is an example enumeration:

X:	4	7	3	2	8	6
Y:	8	10	4	4	9	7

The mean for X is $\hat{\mu}_X = 5$, and the mean for Y is $\hat{\mu}_Y = 7$. Table 17.2 lays out a helpful format for the calculations.

X	Y	$(X-\hat{\mu}_X)$	$(Y-\hat{\mu}_Y)$	$(X-\hat{\mu}_X)^2$	$(Y-\hat{\mu}_Y)^2$	$(X-\hat{\mu}_X)(Y-\hat{\mu}_Y)$
4	8	$4-5=-1$	$8-7=1$	$(-1)^2=1$	$1^2=1$	$(-1)(1)=-1$
7	10	$7-5=2$	$10-7=3$	$2^2=4$	$3^2=9$	$(2)(3)=6$
3	4	$3-5=-2$	$4-7=-3$	$(-2)^2=4$	$(-3)^2=9$	$(-2)(-3)=6$
2	4	$2-5=-3$	$4-7=-3$	$(-3)^2=9$	$(-3)^2=9$	$(-3)(-3)=9$
8	9	$8-5=3$	$9-7=2$	$3^2=9$	$2^2=4$	$(3)(2)=6$
6	7	$6-5=1$	$7-7=0$	$1^2=1$	$(0)^2=1$	$(1)(0)=0$
30	42	0	0	28	32	26

Table 17.2: Covariance calculations.

The bottom numbers are the sums of the columns. The first two are the sums of the enumeration. The next two sums are the sum of the deviations. The next two sums are the sum of the squared deviation. The last is the sum of the cross products. From this table we can find the variance for each variable using the sum of the squared deviations:

$$S_X^2 = \frac{28}{5} = 5.6 \quad S_Y^2 = \frac{32}{5} = 6.4$$

To find the covariance, we need the sum of the cross products:

$$covariance = \frac{\sum_{i=1}^{n}(x_i - \hat{\mu}_X)(y_i - \hat{\mu}_Y)}{n-1} = \frac{26}{6-1} = 5.2$$

Once we have the covariance, we can find the correlation:

$$\hat{r} = \frac{covariance}{\sqrt{S_X^2 S_Y^2}} = \frac{5.2}{\sqrt{(5.6)(6.4)}} = 0.89$$

Note: The only way the correlation is negative is if the covariance is negative. They share the same sign.

17.4 Template

Once it has been determined the test situation is correlatoin, here is how to conduct the test:

H0: The first variable is *not* related to the second variable. H0: $r = 0$.

Alternative hypothesis: One of the following based on the question of interest:

- HA: The first variable is positively related to the second variable. HA: $r > 0$.
- HA: The first variable is negatively related to the second variable. HA: $r < 0$.
- HA: The first variable is related to the second variable. HA: $r \neq 0$.

Equation of the test statistic:

$$t_\delta = \frac{\hat{r}}{\sqrt{\frac{1-\hat{r}^2}{n-2}}}$$

Null distribution: t distribution (Section 9.3.1).

Critical value(s):

1. Find δ = degrees of freedom $= n - 2$.
2. Use Table A.2 to find the CV from the t distribution.
   - If HA: $r > 0$, use the 0.05 column and leave CV positive.
   - If HA: $r < 0$, use the 0.05 column and make CV negative.
   - If HA: $r \neq 0$, use the 0.025 column and use both positive and negative CV.

p-value: The probability statement for the p-value depends on the HA:

- If HA: $r > 0$, p-value $= P(t_\delta > \text{TS})$.
- If HA: $r < 0$, p-value $= P(t_\delta < \text{TS})$.
- If HA: $r \neq 0$, p-value $= 2P(t_\delta > |\text{TS}|)$.

Statistical decision:

- Fail to reject the H0 if p-value $> \alpha$ or
  - if TS < CV and HA: $r > 0$
  - if TS > CV and HA: $r < 0$
  - if TS > negative CV and TS < positive CV and HA: $r \neq 0$
- Reject the H0 if p-value $< \alpha$ or
  - if TS > CV and HA: $r > 0$
  - if TS < CV and HA: $r < 0$
  - if TS < negative CV or TS > positive CV and HA: $r \neq 0$

Conclusion/interpretation:

- If the statistical decision is to fail to reject the H0, then conclude the H0 in English.
- If the statistical decision is to reject the H0, then conclude the HA in English and interpret the estimated correlation value ($\hat{r}$).

17.5 Examples

1. Return to the GRE and GPA scenario (page 174). $\alpha = 0.05$. Results: $\hat{r} = 0.71$.

   Question of interest: Is the GRE quantitative score related to GPA in graduate school?

   Test situation: Correlation

   - One sample: 21 students
   - Two variables: GRE score and graduate GPA
   - Both variables are ratio.

   Hypotheses: Look for key words to find the direction of the relationship, like positive or negative.

   H0: The GRE quantitative score is not related to GPA in graduate school. H0: $r = 0$.

   HA: The GRE quantitative score is related to GPA in graduate school. HA: $r \neq 0$.

   There are no directional words, so the HA is "not equal to."

   Equation of the TS:

   $$t_\delta = \frac{\hat{r}}{\sqrt{\frac{1-\hat{r}^2}{n-2}}}$$

   The t_δ is the null distribution.

   Critical values: Since the equation of the TS has a t_δ, we will use Table A.2: $\delta = n-2 = 21-2 = 19$. Use the 0.025 column in the table, because the HA is a "not equal to." CV $= \pm 2.093$; reject H0 if TS > 2.093 or TS < -2.093.

   Calculate the TS:

   $$TS = \frac{\hat{r}}{\sqrt{\frac{1-\hat{r}^2}{n-2}}} = \frac{0.71}{\sqrt{\frac{1-(0.71)^2}{21-2}}} = 4.39$$

   p-value: The probability statement is p-value $= 2P(t_\delta > |\text{TS}|) = 2P(t_{19} > 4.39)$.

   Statistical decision: Reject the H0 because $4.39(TS) > 2.093(CV)$.

   Conclusion/interpretation: I conclude that the GRE quantitative score is related to GPA in graduate school. Specifically, as GRE score increases, the graduate GPA also increases, or, as GRE score decreases, the graduate GPA also decreases.

2. Return to the coffee and sleep scenario (page 175). $\alpha = 0.05$. Results: $\hat{r} = -0.89$.

Question of interest: The consumption of coffee in the evening before going to bed can be adversely related to your ability to sleep.

Test situation: Correlation

- One sample: 20 graduate students
- Two ratio variables: Amount of coffee drank, hours of sleep

H0: The amount of coffee drank last night is not related to the hours slept last night.

HA: The amount of coffee drank last night is negatively related to the hours slept last night.

HA: $r < 0$. The negatively related comes from the "adversely related" in the question of interest.

Find the CV: $\delta = n - 2 = 20 - 2 = 18$; CV = -1.734; it is negative because the HA is "less than."

Reject the H0 if TS < -1.734.

Calculate the TS:

$$TS = \frac{\hat{r}}{\sqrt{\frac{1-\hat{r}^2}{n-2}}} = \frac{-0.89}{\sqrt{\frac{1-(-0.89)^2}{20-2}}} = -8.28$$

p-value: The probability statement is p-value $= P(t_\delta < \text{TS}) = P(t_{18} < -8.28)$.

Statistical decision: Reject the H0 because $-8.28(TS) < -1.734(CV)$.

Conclusion/interpretation: I conclude that the amount of coffee drank last night is negatively related to the hours slept last night. Specifically, as the amount of coffee drank last night increases, the hours slept decreases, or, as the amount of coffee drank last night decreases, the hours slept increases.

3. Do larger homes tend to cost more? To answer this question, suppose 82 homes that are for sale were measured for their square footage and selling price. $\alpha = 0.05$. Results: $\hat{r} = 0.45$.

Question of interest: Do larger homes tend to cost more?

Test situation: Correlation

- One sample: 82 homes that are for sale
- Two ratio variables: Home's square footage and home's selling price

H0: A home's square footage is not related to the home's selling price.

HA: A home's square footage is positively related to a home's selling price. HA: $r > 0$.

The question of interest indicates larger homes and costing more. This is a positive relationship.

Find the CV: CV $= 1.664$; $\delta = n - 2 = 82 - 2 = 80$. Reject the H0 if $TS > 1.664$.

Calculate the TS:

$$TS = \frac{\hat{r}}{\sqrt{\frac{1-\hat{r}^2}{n-2}}} = \frac{0.45}{\sqrt{\frac{1-(0.45)^2}{82-2}}} = 4.51$$

p-value: The probability statement is p-value $= P(t_\delta > \text{TS}) = P(t_{80} > 4.51)$.

Statistical decision: Reject the H0 because 4.51(TS) > 1.664(CV).

Conclusion/interpretation: I conclude that a home's square footage is positively related to the home's selling price. Specifically, as a home's square footage increases, the selling price also increases.

4. Are ice cream sales related to birth rate? To answer this question, suppose 17 randomly selected U.S. cities had records of ice cream sales in dollars and number of births. Results: $\hat{r} = 0.32$; $\alpha = 0.05$.

Question of interest: Are ice cream sales related to birth rate?

Test situation: Correlation

- One sample: 17 U.S. cities
- Two ratio variables: Ice cream sales and number of births

H0: Ice cream sales are not related to the number of births.

HA: Ice cream sales are related to the number of births.

Find the CV:

(a) $\delta = n - 2 = 17 - 2 = 15$

(b) CV $= \pm 2.131$

Calculate the TS:

$$TS = \frac{\hat{r}}{\sqrt{\frac{1-\hat{r}^2}{n-2}}} = \frac{.32}{\sqrt{\frac{1-(.32)^2}{17-2}}} = 1.31$$

p-value: The probability statement is p-value $= 2P(t_\delta > |\text{TS}|) = P(t_{15} > 1.31)$.

Statistical decision: Fail to reject the H0 because 1.31(TS) is not greater than 2.131(CV).

Conclusion/interpretation: I conclude that ice cream sales are not related to the number of births.

17.6 Correlation Versus Causation

If two variables are correlated to each other does, that mean one variable causes the variable? The answer is *no.* The reason people think correlation could be causation is they try and justify it. Let's look at some examples:

1. Ice cream sales are related to the number of births.

   Even if the two variables are highly correlated, it doesn't make any sense that ice cream sales cause births.

2. Home size is related to home price.

   It may be true that home size causes home price, but correlation does not assess that causation. Correlation only assessed the linear pattern.

3. For 7 years in the 1930s, a biologist measured the number of storks and the population in Oldenburg, Germany.[1] The correlation is $\hat{r} = 0.97$, so as the number of storks increased, the population in Oldenburg increased. Are storks bringing babies? Absolutely not! So correlation is not causation.

Correlation and causation are a unidirectional relationship:

- If variable 1 causes variable 2, then variable 1 and variable 2 should be correlated.
- If variable 1 is correlated to variable 2, it does *not* imply that one caused the other.

If two variables are correlated we do not know what caused what.

17.7 Examples With R

1. Test if soduim and fat are related (page 173):

   HA: Soduim and fat are related. HA: $r \neq 0$.

   ```
   > cor(sodium, fat) # correlation value
   [1] -0.1154668
   > cor.test(sodium, fat, alternative='two.sided')
   ```

[1] Box, G. E. P., Hunter, J. S., & Hunter, W. G. (2005). *Statistics for Experimenters: Design, Innovation, and Discovery* (2nd ed.). Wiley-Interscience, p. 8.

```
Pearson's product-moment correlation

data:  sodium and fat
t = -0.25993, df = 5, p-value = 0.8053
alternative hypothesis: true correlation is not equal to 0
95 percent confidence interval:
 -0.7990454  0.6983117
sample estimates:
       cor
-0.1154668
```

The "t" is the test statistic value. The "df" is the degrees of freedom, or δ. The p-value is greater than 0.05; thus, fail to reject the H0. The "cor" is the correlation value, or $\hat{r}$. The interpretation is to conclude that sodium is not related to fat in popular fast food meals.

2. Return to the GRE and GPA scenario (page 182):

```
> r <- 0.71
> n <- 21
> TS <- r/sqrt((1-r^2)/(n-2))
> pt(abs(TS), n-2, lower.tail=FALSE)*2
[1] 0.0003113873
```

The p-value is less than 0.05; thus, the statistical decision is to reject the H0.

3. Return to the coffee and sleep scenario (page 182):

```
> r <- -0.89
> n <- 20
> TS <- r/sqrt((1-r^2)/(n-2))
> pt(abs(TS), n-2, lower.tail=FALSE)*2
[1] 7.464563e-08
```

The p-value is less than 0.05; thus, the statistical decision is to reject the H0.

Chapter 18

Independence

The independence test situation has one sample and two categorical variables. The purpose is similar to the correlation test situation: to compare the variables. Here the variables have a categorical level of measurement.

18.1 Terms

Independence test: One samples with two categorical variables.

O: Observed frequency.

E: Expected frequency.

Contingency table: A table with two or more rows and two or more columns (Section 15.2). The boxes formed by the table are called cells. The numbers in the cells are frequencies.

Row marginal sum: The sum of the frequencies in a particular row.

Column marginal sum: The sum of the frequencies in a particular column.

Total sum: The sum of all the frequencies, or the sum of all the row marginal sums, or the sum of all the column marginal sums.

The independence test situation also uses a contingency table, like in homogeneity. The only difference is variable 1 outcomes are in the rows, and variable 2 outcomes are in the columns. See Section 15.2 for a review of contingency tables.

18.2 Template

Once it has been determined that the test situation is independence, here is how to conduct the test:

H0: The first variable is not related to the second variable. Or, the first variable is independent of the second variable.

HA: The first variable is related to the second variable.

Equation of the statistic:

$$\chi^2_\delta = sum\frac{(O-E)^2}{E}$$

Null distribution: χ^2 distribution (Section 9.3.2)

Critical value:

1. δ = degrees of freedom = (# row -1)(# column -1).
2. Use the 0.05 column (since $\alpha = 0.05$) in Table A.1 to find the CV from the χ^2 distribution.

p-value: The probability statement is p-value $= P(\chi^2_\delta > \text{TS})$.

Calculate the TS:

1. E = (row total)(column total)/(total sum)
2. Calculate the $O - E$ table
3. Calculate $(O - E)^2$ table
4. Calculate $(O - E)^2/E$ table
5. TS = sum$(O - E)^2/E$ table

Statistical decision:

- Fail to reject the H0 if TS $<$ CV or if p-value $> \alpha$.
- Reject the H0 if TS $>$ CV or if p-value $< \alpha$.

Conclusion/interpretation:

- If the statistical decision is to fail to reject the H0, then conclude the H0 in English.
- If the statistical decision is to reject the H0, then conclude the HA in English and interpret the two largest O - E values.

18.3 Examples

1. Are movie genres related to school subjects? Suppose 200 seventh graders were asked the following two questions:

   - What is your favorite movie genre: horror, romance, or action?
   - What is your favorite subject in school: English, math, science, or art?

   $\alpha = 0.05$. Results:

	English	Math	Science	Art
Horror	10	13	15	12
Romance	19	20	21	10
Action	11	47	14	8

Question of interest: Are movie genres related to school subjects?

Test situation: Independence

- One sample: 200 seventh graders
- Two categorical variables: Favorite movie genre and favorite subject in school

H0: Movie genre is not related to school subjects.

HA: Movie genre is related to school subjects.

Equation of test statistic:

$$\chi^2_\delta - sum\frac{(O-E)^2}{E}$$

The null distribution is χ^2_δ.

Find the CV: CV $= 12.592$; $\delta =$ (number of rows -1)(number of columns -1) $= (3-1)(4-1) = 6$.

Reject the H0 if TS > 12.592

Calculate the TS: The calculation of the TS is exactly the same as in homogeneity.

O	English	Math	Science	Art	
Horror	10	13	15	12	50
Romance	19	20	21	10	70
Action	11	47	14	8	80
	40	80	50	30	

E	English	Math	Science	Art
Horror	10	20	12.5	7.5
Romance	14	28	17.5	10.5
Action	16	32	20.0	12.0

$O-E$	English	Math	Science	Art
Horror	0	-7	2.5	4.5
Romance	5	-8	3.5	-0.5
Action	-5	15	-6.0	-4.0

$(O-E)^2/E$	English	Math	Science	Art
Horror	0.00	2.45	0.50	2.70
Romance	1.79	2.29	0.70	0.02
Action	1.56	7.03	1.80	1.33

$$\text{TS} = 0.00 + 2.45 + 0.50 + 2.70 + 1.79 + 2.29 + 0.70 + 0.02 + 1.56 + 7.03 + 1.80 + 1.33 = 22.17$$

p-value: The probability statement is p-value $= P(\chi^2_\delta > \text{TS}) = P(\chi^2_6 > 22.17)$.

Statistical decision: Reject the H0 because 22.17(TS) > 12.592(CV).

Conclusion/Interpretation: I conclude that movie genres are related to school subjects. Specifically, seventh graders who like math also like action movies (largest O - E), and seventh graders who like English also like romance movies (second largest O - E).

2. Is your belief about climate change (global warming) related to where you live? To answer this question, suppose 100 people were asked the following: "Do you believe climate change (global warming) is a serious problem facing the United States: yes or no?" and "Where do you live: East Coast, West Coast, central states, or Rocky Mountain states?" $\alpha = 0.05$. Results:

	Yes	No
East Coast	12	8
West Coast	15	15
Central states	6	14
Rocky Mt. states	7	23

Question of interest: Is your belief about climate change (global warming) related to where you live?

Test situation: Independence

- One sample: 100 people
- Two categorical variables: Belief in climate change and where do you live

H0: Belief about climate change is not related to where you live.

HA: Belief about climate change is related to where you live.

Find the CV: CV = 7.815; $\delta = (4-1)(2-1) = 3$. Reject the H0 if TS > 7.815.

Calculate the TS: Create each of the tables:

E	Yes	No
East Coast	8	12
West Coast	12	18
Central states	8	12
Rocky Mt. states	12	18

$O - E$	Yes	No
East Coast	4	-4
West Coast	3	-3
Central states	-2	2
Rocky Mt. states	-5	5

$(O-E)^2/E$	Yes	No
East Coast	2.00	1.33
West Coast	0.75	0.50
Central states	0.50	0.33
Rocky Mt. states	2.08	1.39

TS $= 2 + 0.75 + 0.5 + 2.08 + 1.33 + 0.5 + 0.33 + 1.39 = 8.89$

p-value: The probability statement is p-value $= P(\chi^2_\delta > \text{TS}) = P(\chi^2_3 > 8.89)$.

Statistical decision: Reject H0 because 8.89(TS) > 7.81(CV).

Conclusion/interpretation: I conclude that belief in climate change is related to where you live. Specifically, people who live in the Rocky Mountain states do not believe in climate change (largest O - E), and people who live on the East Coast believe in climate change (second largest O - E).

3. Is gender related to wearing a costume on Halloween? Suppose 200 college students were asked, "What is your gender: female or male?" and "Last Halloween did you wear a costume: yes or no?" $\alpha = 0.05$. Results:

	Yes	No
Female	59	46
Male	51	44

Question of interest: Is gender related to wearing a costume on Halloween?

Test situation: Independence

- One sample: 200 college students
- Two categorical variables: Gender and wearing a costume

H0: Gender is not related to wearing a costume on Halloween.

HA: Gender is related to wearing a costume on Halloween.

Find the CV: CV $= 3.84$; $\delta = (2-1)(2-1) = 1$.

Calculate the TS: TS $= 0.13$ (calculations not shown)

p-value: The probability statement is p-value $= P(\chi^2_\delta > \text{TS}) = P(\chi^2_1 > 0.13)$.

Statistical decision: Fail to reject H0 because 0.13(TS) < 3.84(CV).

Conclusion/interpretation: Gender is not related to wearing a costume on Halloween.

18.4 Examples With R

1. Return to the movie genre scenario (page 189):

```
> chisq.test(Omovie, correct=FALSE)

Pearson's Chi-squared test


data:  Omovie

X-squared = 22.172, df = 6, p-value = 0.001127

> Omovie-chisq.test(Omovie)$expected

         English Math Science  Art

Horror         0   -7     2.5  4.5

Romance        5   -8     3.5 -0.5

Action        -5   15    -6.0 -4.0
```

The "X-squared" is the test statistic. The "df" is the degrees of freedom or, δ. The p-value is less than 0.05; thus, the same statistical decision. The $O - E$ table is provided.

2. Return to the climate change scenario (page 190):

```
> chisq.test(Oclimate, correct=FALSE)

Pearson's Chi-squared test


data:  Oclimate

X-squared = 8.8889, df = 3, p-value = 0.03081

> Oclimate-chisq.test(Oclimate)$expected

                 Yes No

East Coast         4 -4

West Coast         3 -3

Central states    -2  2

Rocky Mt. states  -5  5
```

The "X-squared" is the test statistic. The "df" is the degrees of freedom, or δ. The p-value is less than 0.05; thus, the same statistical decision. The $O - E$ table is provided.

3. Return to the costume scenario (page 191):

```
> chisq.test(Ocostume, correct=FALSE)

Pearson's Chi-squared test


data:  Ocostume
X-squared = 0.12658, df = 1, p-value = 0.722
```

The "X-squared" is the test statistic. The "df" is the degrees of freedom, or δ. The p-value is greater than 0.05; thus, we reach the same statistical decision. The $O - E$ table is not provided because it is not required based on the statistical decision.

Appendix A

Critical Values

δ	0.1	0.05	0.025	0.01
1	2.706	3.841	5.024	6.635
2	4.605	5.991	7.378	9.210
3	6.251	7.815	9.348	11.345
4	7.779	9.488	11.143	13.277
5	9.236	11.070	12.833	15.086
6	10.645	12.592	14.449	16.812
7	12.017	14.067	16.013	18.475
8	13.362	15.507	17.535	20.090
9	14.684	16.919	19.023	21.666
10	15.987	18.307	20.483	23.209
11	17.275	19.675	21.920	24.725
12	18.549	21.026	23.337	26.217
13	19.812	22.362	24.736	27.688
14	21.064	23.685	26.119	29.141
15	22.307	24.996	27.488	30.578
16	23.542	26.296	28.845	32.000
17	24.769	27.587	30.191	33.409
18	25.989	28.869	31.526	34.805
19	27.204	30.144	32.852	36.191
20	28.412	31.410	34.170	37.566

χ^2 Table

Table A.1: Provides x values such that $P(\chi^2_\delta > x) = p$ for the column values of p.

t Table

δ	0.1	0.05	0.025	0.01
1	3.078	6.314	12.706	31.821
2	1.886	2.920	4.303	6.965
3	1.638	2.353	3.182	4.541
4	1.533	2.132	2.776	3.747
5	1.476	2.015	2.571	3.365
6	1.440	1.943	2.447	3.143
7	1.415	1.895	2.365	2.998
8	1.397	1.860	2.306	2.896
9	1.383	1.833	2.262	2.821
10	1.372	1.812	2.228	2.764
11	1.363	1.796	2.201	2.718
12	1.356	1.782	2.179	2.681
13	1.350	1.771	2.160	2.650
14	1.345	1.761	2.145	2.624
15	1.341	1.753	2.131	2.602
16	1.337	1.746	2.120	2.583
17	1.333	1.740	2.110	2.567
18	1.330	1.734	2.101	2.552
19	1.328	1.729	2.093	2.539
20	1.325	1.725	2.086	2.528
21	1.323	1.721	2.080	2.518
22	1.321	1.717	2.074	2.508
23	1.319	1.714	2.069	2.500
24	1.318	1.711	2.064	2.492
25	1.316	1.708	2.060	2.485
26	1.315	1.706	2.056	2.479
27	1.314	1.703	2.052	2.473
28	1.313	1.701	2.048	2.467
29	1.311	1.699	2.045	2.462
30	1.310	1.697	2.042	2.457
31	1.309	1.696	2.040	2.453
32	1.309	1.694	2.037	2.449
33	1.308	1.692	2.035	2.445
34	1.307	1.691	2.032	2.441
35	1.306	1.690	2.030	2.438
40	1.303	1.684	2.021	2.423
50	1.299	1.676	2.009	2.403
60	1.296	1.671	2.000	2.390
70	1.294	1.667	1.994	2.381
80	1.292	1.664	1.990	2.374
100	1.290	1.660	1.984	2.364
∞	1.282	1.645	1.960	2.326

Table A.2: Provides x values such that $P(t_\delta > x) = p$ for the column values of p.

Appendix B

Standard Normal Probabilities

z	0	0.01	0.02	0.03	0.04	0.05	0.06	0.07	0.08	0.09
	Standard Normal Probability Table $P(Z < z)$									
-0.0	0.50000	0.49601	0.49202	0.48803	0.48405	0.48006	0.47608	0.47210	0.46812	0.46414
-0.1	0.46017	0.45620	0.45224	0.44828	0.44433	0.44038	0.43644	0.43251	0.42858	0.42465
-0.2	0.42074	0.41683	0.41294	0.40905	0.40517	0.40129	0.39743	0.39358	0.38974	0.38591
-0.3	0.38209	0.37828	0.37448	0.37070	0.36693	0.36317	0.35942	0.35569	0.35197	0.34827
-0.4	0.34458	0.34090	0.33724	0.33360	0.32997	0.32636	0.32276	0.31918	0.31561	0.31207
-0.5	0.30854	0.30503	0.30153	0.29806	0.29460	0.29116	0.28774	0.28434	0.28096	0.27760
-0.6	0.27425	0.27093	0.26763	0.26435	0.26109	0.25785	0.25463	0.25143	0.24825	0.24510
-0.7	0.24196	0.23885	0.23576	0.23270	0.22965	0.22663	0.22363	0.22065	0.21770	0.21476
-0.8	0.21186	0.20897	0.20611	0.20327	0.20045	0.19766	0.19489	0.19215	0.18943	0.18673
-0.9	0.18406	0.18141	0.17879	0.17619	0.17361	0.17106	0.16853	0.16602	0.16354	0.16109
-1.0	0.15866	0.15625	0.15386	0.15151	0.14917	0.14686	0.14457	0.14231	0.14007	0.13786
-1.1	0.13567	0.13350	0.13136	0.12924	0.12714	0.12507	0.12302	0.12100	0.11900	0.11702
-1.2	0.11507	0.11314	0.11123	0.10935	0.10749	0.10565	0.10383	0.10204	0.10027	0.09853
-1.3	0.09680	0.09510	0.09342	0.09176	0.09012	0.08851	0.08691	0.08534	0.08379	0.08226
-1.4	0.08076	0.07927	0.07780	0.07636	0.07493	0.07353	0.07215	0.07078	0.06944	0.06811
-1.5	0.06681	0.06552	0.06426	0.06301	0.06178	0.06057	0.05938	0.05821	0.05705	0.05592
-1.6	0.05480	0.05370	0.05262	0.05155	0.05050	0.04947	0.04846	0.04746	0.04648	0.04551
-1.7	0.04457	0.04363	0.04272	0.04182	0.04093	0.04006	0.03920	0.03836	0.03754	0.03673
-1.8	0.03593	0.03515	0.03438	0.03362	0.03288	0.03216	0.03144	0.03074	0.03005	0.02938
-1.9	0.02872	0.02807	0.02743	0.02680	0.02619	0.02559	0.02500	0.02442	0.02385	0.02330
-2.0	0.02275	0.02222	0.02169	0.02118	0.02068	0.02018	0.01970	0.01923	0.01876	0.01831
-2.1	0.01786	0.01743	0.01700	0.01659	0.01618	0.01578	0.01539	0.01500	0.01463	0.01426
-2.2	0.01390	0.01355	0.01321	0.01287	0.01255	0.01222	0.01191	0.01160	0.01130	0.01101
-2.3	0.01072	0.01044	0.01017	0.00990	0.00964	0.00939	0.00914	0.00889	0.00866	0.00842
-2.4	0.00820	0.00798	0.00776	0.00755	0.00734	0.00714	0.00695	0.00676	0.00657	0.00639
-2.5	0.00621	0.00604	0.00587	0.00570	0.00554	0.00539	0.00523	0.00508	0.00494	0.00480
-2.6	0.00466	0.00453	0.00440	0.00427	0.00415	0.00402	0.00391	0.00379	0.00368	0.00357
-2.7	0.00347	0.00336	0.00326	0.00317	0.00307	0.00298	0.00289	0.00280	0.00272	0.00264
-2.8	0.00256	0.00248	0.00240	0.00233	0.00226	0.00219	0.00212	0.00205	0.00199	0.00193
-2.9	0.00187	0.00181	0.00175	0.00169	0.00164	0.00159	0.00154	0.00149	0.00144	0.00139
-3.0	0.00135	0.00131	0.00126	0.00122	0.00118	0.00114	0.00111	0.00107	0.00104	0.00100
-3.1	0.00097	0.00094	0.00090	0.00087	0.00084	0.00082	0.00079	0.00076	0.00074	0.00071
-3.2	0.00069	0.00066	0.00064	0.00062	0.00060	0.00058	0.00056	0.00054	0.00052	0.00050
-3.3	0.00048	0.00047	0.00045	0.00043	0.00042	0.00040	0.00039	0.00038	0.00036	0.00035
-3.4	0.00034	0.00032	0.00031	0.00030	0.00029	0.00028	0.00027	0.00026	0.00025	0.00024
-3.5	0.00023	0.00022	0.00022	0.00021	0.00020	0.00019	0.00019	0.00018	0.00017	0.00017
-3.6	0.00016	0.00015	0.00015	0.00014	0.00014	0.00013	0.00013	0.00012	0.00012	0.00011
-3.7	0.00011	0.00010	0.00010	0.00010	0.00009	0.00009	0.00008	0.00008	0.00008	0.00008
-3.8	0.00007	0.00007	0.00007	0.00006	0.00006	0.00006	0.00006	0.00005	0.00005	0.00005
-3.9	0.00005	0.00005	0.00004	0.00004	0.00004	0.00004	0.00004	0.00004	0.00003	0.00003

Table B.1: Standard normal distribution probabilities for less than negative z-scores.

Standard Normal Probability Table $P(Z < z)$

z	0	0.01	0.02	0.03	0.04	0.05	0.06	0.07	0.08	0.09
0.0	0.50000	0.50399	0.50798	0.51197	0.51595	0.51994	0.52392	0.52790	0.53188	0.53586
0.1	0.53983	0.54380	0.54776	0.55172	0.55567	0.55962	0.56356	0.56749	0.57142	0.57535
0.2	0.57926	0.58317	0.58706	0.59095	0.59483	0.59871	0.60257	0.60642	0.61026	0.61409
0.3	0.61791	0.62172	0.62552	0.62930	0.63307	0.63683	0.64058	0.64431	0.64803	0.65173
0.4	0.65542	0.65910	0.66276	0.66640	0.67003	0.67364	0.67724	0.68082	0.68439	0.68793
0.5	0.69146	0.69497	0.69847	0.70194	0.70540	0.70884	0.71226	0.71566	0.71904	0.72240
0.6	0.72575	0.72907	0.73237	0.73565	0.73891	0.74215	0.74537	0.74857	0.75175	0.75490
0.7	0.75804	0.76115	0.76424	0.76730	0.77035	0.77337	0.77637	0.77935	0.78230	0.78524
0.8	0.78814	0.79103	0.79389	0.79673	0.79955	0.80234	0.80511	0.80785	0.81057	0.81327
0.9	0.81594	0.81859	0.82121	0.82381	0.82639	0.82894	0.83147	0.83398	0.83646	0.83891
1.0	0.84134	0.84375	0.84614	0.84849	0.85083	0.85314	0.85543	0.85769	0.85993	0.86214
1.1	0.86433	0.86650	0.86864	0.87076	0.87286	0.87493	0.87698	0.87900	0.88100	0.88298
1.2	0.88493	0.88686	0.88877	0.89065	0.89251	0.89435	0.89617	0.89796	0.89973	0.90147
1.3	0.90320	0.90490	0.90658	0.90824	0.90988	0.91149	0.91309	0.91466	0.91621	0.91774
1.4	0.91924	0.92073	0.92220	0.92364	0.92507	0.92647	0.92785	0.92922	0.93056	0.93189
1.5	0.93319	0.93448	0.93574	0.93699	0.93822	0.93943	0.94062	0.94179	0.94295	0.94408
1.6	0.94520	0.94630	0.94738	0.94845	0.94950	0.95053	0.95154	0.95254	0.95352	0.95449
1.7	0.95543	0.95637	0.95728	0.95818	0.95907	0.95994	0.96080	0.96164	0.96246	0.96327
1.8	0.96407	0.96485	0.96562	0.96638	0.96712	0.96784	0.96856	0.96926	0.96995	0.97062
1.9	0.97128	0.97193	0.97257	0.97320	0.97381	0.97441	0.97500	0.97558	0.97615	0.97670
2.0	0.97725	0.97778	0.97831	0.97882	0.97932	0.97982	0.98030	0.98077	0.98124	0.98169
2.1	0.98214	0.98257	0.98300	0.98341	0.98382	0.98422	0.98461	0.98500	0.98537	0.98574
2.2	0.98610	0.98645	0.98679	0.98713	0.98745	0.98778	0.98809	0.98840	0.98870	0.98899
2.3	0.98928	0.98956	0.98983	0.99010	0.99036	0.99061	0.99086	0.99111	0.99134	0.99158
2.4	0.99180	0.99202	0.99224	0.99245	0.99266	0.99286	0.99305	0.99324	0.99343	0.99361
2.5	0.99379	0.99396	0.99413	0.99430	0.99446	0.99461	0.99477	0.99492	0.99506	0.99520
2.6	0.99534	0.99547	0.99560	0.99573	0.99585	0.99598	0.99609	0.99621	0.99632	0.99643
2.7	0.99653	0.99664	0.99674	0.99683	0.99693	0.99702	0.99711	0.99720	0.99728	0.99736
2.8	0.99744	0.99752	0.99760	0.99767	0.99774	0.99781	0.99788	0.99795	0.99801	0.99807
2.9	0.99813	0.99819	0.99825	0.99831	0.99836	0.99841	0.99846	0.99851	0.99856	0.99861
3.0	0.99865	0.99869	0.99874	0.99878	0.99882	0.99886	0.99889	0.99893	0.99896	0.99900
3.1	0.99903	0.99906	0.99910	0.99913	0.99916	0.99918	0.99921	0.99924	0.99926	0.99929
3.2	0.99931	0.99934	0.99936	0.99938	0.99940	0.99942	0.99944	0.99946	0.99948	0.99950
3.3	0.99952	0.99953	0.99955	0.99957	0.99958	0.99960	0.99961	0.99962	0.99964	0.99965
3.4	0.99966	0.99968	0.99969	0.99970	0.99971	0.99972	0.99973	0.99974	0.99975	0.99976
3.5	0.99977	0.99978	0.99978	0.99979	0.99980	0.99981	0.99981	0.99982	0.99983	0.99983
3.6	0.99984	0.99985	0.99985	0.99986	0.99986	0.99987	0.99987	0.99988	0.99988	0.99989
3.7	0.99989	0.99990	0.99990	0.99990	0.99991	0.99991	0.99992	0.99992	0.99992	0.99992
3.8	0.99993	0.99993	0.99993	0.99994	0.99994	0.99994	0.99994	0.99995	0.99995	0.99995
3.9	0.99995	0.99995	0.99996	0.99996	0.99996	0.99996	0.99996	0.99996	0.99997	0.99997

Table B.2: Standard normal distribution probabilities for less than positive z-scores.

Printed in the USA
CPSIA information can be obtained
at www.ICGtesting.com
LVHW080615040624
782165LV00024B/69